Medicinal Plants

Chemistry and Properties

Medicinal Plants

Chemistry and Properties

M. Daniel

SCIENCE PUBLISHERS
An imprint of Edenbridge Ltd., British Channel Islands.
Post Office Box 699
Enfield, New Hampshire 03748
United States of America

Internet site: *http://www.scipub.net*

sales@scipub.net (marketing department)
editor@scipub.net (editorial department)
info@scipub.net (for all other enquiries)

Library of Congress Cataloging-in-Publication Data

Daniel, M., Dr.
Medicinal plants : chemistry and properties / M. Daniel.
p. cm.
Includes bibliographical references (p.) and index.
ISBN 1-57808-395-8
1. Medicinal plants. 2. Materia medica, Vegetable. 1. Title.

QK99.AID36 2005
615'.321--dc22

2005054084

ISBN 1-57808-395-8

Published by Science Publishers, Enfield, NH, USA
An imprint of Edenbridge Ltd.
Printed in India.

Preface

A thorough knowledge of the chemistry of medicinal plants is essential in the WTO regime to understand the manifold activities of the plants, the quality of raw materials, extracts and formulations and also to market the plant-derived drugs. Unfortunately, this aspect is not given its due importance in most of the treatises of medicinal herbs. Of late, there has emerged a great interest in the constituents of medicinal plants. This is amply evident by the recent spurt of publications in *Journal of Natural Products, Phytochemistry* and a large number of journals on chemistry. Even then, a concise account of plant products remains out of the reach of a student or a researcher who would like to initiate work on these fascinating groups of compounds.

Unfortunately, such texts, of an interdisciplinary nature, are not available, and if available are not within the reach of common man. I have tried to explain the chemistry of useful plants in 1990, in a book *A Phytochemical Approach to Economic Botany.* In this treatise, a general account of plant products (both primary and secondary metabolites) and the utility of individual plants were explained concisely. I was able to include only about 100 medicinal plants. In the ensuing 15 years, great strides were achieved by natural product chemistry all over the world. More and more new plants of varied properties and diverse uses for known plants were discovered mainly due to the leads taken from ethnobotany. Data on the chemical constituents on important plants increased by leaps and bounds. Experimental proof on the activities of individual compounds was also made available. All these data need to be assorted and arranged properly in a user-friendly form. My book is a step in this direction.

Since each and every plant in nature exhibits one or other medicinal property and every second plant is considered an important source of medicine, the selection of plants for this book was a daunting task. India alone boasts of more than 2000 medicinal plants, while the total number of medicinal herbs in the world is about 70,000. Among the 350 herbs I have selected for this book, 110 medicinal herbs belong to Europe and America inclusive of the bestselling herbs of the USA (Ginseng, *Ginkgo*, Garlic, *Echinacea* and Goldenseal), around 200 plants of Indian scene and a good number of plants from S.E. Asia and China. The selection was guided by the importance of the plants in world market and representation of members of different taxonomic groups (say families) as well as of different classes (and subclasses) of compounds. All the available data on the chemical compounds and the pharmacological studies on these plants/compounds have been incorporated. I have reduced the bibliography tremendously in the sense that chemical reports published before 1980 (for which many books are available) were cited rarely and a judicious selection of recent literature is resorted to. For the collection of data on Indian plants, I have relied heavily on the *Wealth of India* volumes published by CSIR, India. As far as possible, latest information is given. (I do not know how this is possible. Even when I pen this, new compounds are being added from the same or related plants!)

The plants are arranged on the type and nature of chemical compounds they contain. I faced difficulty with a number of plants like *Ginkgo*, which happen to be rich sources of more than one type of compounds. In such cases, I opted to include them on the basis of compounds with proven pharmacological activity. There are also plants like *Boerhavia*, found to contain more than two types of compounds possessing different properties. For example, punarnavoside, a phenolic glycoside from this plant, is an antifibrinolytic agent occurring along with hypoxanthine, which is hypotensive. In such cases I had to strike a compromise.

I have added secondary metabolites to this volume firstly due to their great therapeutic role. (By this, I am not, in anyway, underestimating the role of primary metabolites in healing process). Among the secondary metabolites, alkaloids are listed at the top followed by terpenoids. Phenolics, with their highlighted role as antioxidants and in the maintenance of all the systems of the body—they contain surprisingly large numbers of plants—are placed third. I prefer to keep gums and mucilages along with secondary metabolites even though they are carbohydrates. The chapter on primary metabolites is a small one, but I am sure more and more plants will be added in this category in course of time.

Another grouping of medicinal herbs based on their medicinal properties (a user's guide) and some other useful handy information have been appended in the end.

The help rendered by Mr Jinu Skariah in typing the manuscript is gratefully acknowledged.

M. Daniel
Department of Botany,
The M.S. University of Baroda,
Vadodara, INDIA

Contents

Preface v

1. Introduction 1

PART A: SECONDARY METABOLITES

2. Alkaloids 9

2.1 Alkaloidal amines 11
Ephedra sinica Stapf. (Ephedraceae) 11
Alhagi pseudalhagi Desv. (Fabaceae) 12
Catha edulis Forsk (Celastraceae) 13
Sida cordifolia Linn. (Malvaceae) 13
Urtica dioica Linn. (Urticaceae) 13

2.2 Diterpenoid Alkaloids 14
Aconitum napellus Linn. (Ranunculaceae) 14
Aconitum ferox Wall 15
Aconitum heterophyllum Wall 15
Aconitum chasmantham Stapf. 15

2.3 Imidazole Alkaloids 16
Pilocarpus spp. (Rutaceae) 16

2.4 Indole Alkaloids 16
Catharanthus roseus G. Don (Apocynaceae) 17
Uncaria guianensis Gmeliu *& U. tomentosa* DC. (Rubiaceae) 18
Strychnos nux-vomica Linn. (Loganiaceae) 18
Physostigma venenosum Bal. (Fabaceae) 19
Mucuna pruriens DC. (Fabaceae) 19
Claviceps purpurea Tul. (Clavicipitaceae) 20
Convolvulus pluricaulis Chois. (Convolvulaceae) 21
Evolvulus alsinoides Linn. (Convolvulaceae) 21
Ipomoea nil Roth. (Convolvulaceae) 21
Ipomoea mauritiana Jacq. (Convolvulaceae) 22
Ipomoea marginata Verdc. (Convolvulaceae) 22
Erythrina variegata Linn. (Fabaceae) 23
Rauwoflia serpentina Benth (Apocynaceae) 23
Passiflora incarnata Linn. (Passifloraceae) 24
Alstonia scholaris R. Br. (Apocynaceae) 24
Peganum harmala Linn. (Zygophyllaceae) 25
Desmodium gangeticum DC. (Fabaceae) 25

Murraya exotica Blance (Rutaceae) 26
Gelsemium sempervirens Ait. (Loganiaceae) 26
Tabernaemontana divaricata Roem. & Schult. (Apocynaceae) 26
2.5 Isoquinoline Alkaloids 27
Cephaelis ipecacuanha A. Rich. (Rubiaceae) 28
Berberis aristata DC (Berberidaceae) 29
Alangium salvifolium Wang (Alangiaceae) 29
Hydrastis canadensis Linn. (Ranunculaceae) 30
Mahonia aquifolium Nuttal (Berberidaceae) 31
Argemone mexicana Linn. (Papaveraceae) 31
Sanguinaria canadensis Linn. (Papaveraceae) 31
Curare 32
Tinospora cordifolia Miers (Menispermaceae) 32
Cissampelos pariera Linn. (Menispermaceae) 33
Stephania glabra Miers. (Menispermaceae) 33
Jateorhiza palmata Linn. (Menispermaceae) 34
Tiliacora racemosa Cole. (Menispermaceae) 34
Papaver somniferum Linn. (Papaveraceae) 34
Aristolochia indica Linn. (Aristolochiaceae) 35
Hedera nepalensis Koch. (Araliaceae) 36
Nelumbo nucifera Gaertn (Nymphaeaceae) 36
2.6 Monoterpenoid Alkaloids 37
Valeriana officinalis Linn. (Valerianaceae) 37
Enicostemma hyssopifolium Verdoon (Gentianaceae) 37
2.7 Phenanthro–indolizidine Alkaloids 37
Tylophora indica Merrill. (Asclepiadaceae) 38
2.8 Pyridine-piperidine Alkaloids 38
Areca catechu Linn. (Arecaceae) 38
Piper nigrum Linn. (Piperaceae) 39
Piper longum Linn 39
Punica granatum Linn. (Punicaceae) 40
Conium maculatum Linn. (Apiaceae) 40
Lobelia inflata Linn. (Lobeliaceae) 40
2.9 Quinazoline Alkaloids 41
Adhatoda zeylanica Medic. (Acanthaceae) 41
2.10 Quinoline Alkaloids 41
Cinchona spp. (Rubiaceae) 42
Toddalia asiatica Lam. (Rutaceae) 42
Glycosmis pentaphylla DC (Rutaceae) 43
2.11 Quinolizidines/Lupinane Alkaloids 43
Cytisus scoparius Linn. (Fabaceae) 43
Boerhavia diffusa Linn. (Nyctaginaceae) 44
2.12 Sesquiterpene Alkaloids 44
Celastrus paniculatus Willd (Celastraceae) 44
2.13 Steroidal Alkaloids 45
Veratrum viride Aiton (Liliaceae) 45

Holarrhena pubescens Wall. (Apocynaceae) 46
Solanum americanum Linn. (Solanaceae) 46
Solanum surattense Burm. f. (Euphorbiaceae) 47
Wrightia arborea Mabb. (Apocynaceae) 47
2.14 Tropane Alkaloids 48
Atropa belladonna Linn. (Solanaceae) 48
Hyoscyamus niger Linn. (Solanaceae) 49
Datura stamonium Linn. (Solanaceae) 49
Datura metel Linn. 50
Duboisia myoporoides R. Br. & D. *leichhardtii* F. Muell. (Solanaceae) 50
Scopolia carniolica Jacq. (Solanaceae) 50
Mandragora officinarum Linn. (Solanaceae) 50
Erythroxylum coca Lam. (Erythroxylaceae) 51
Convolvulus arvensis Linn (Convolvulaceae) 51
2.15 Tropolone Alkaloids 52
Colchicum autumnale Linn. (Liliaceae) 52
Gloriosa superba Linn. (Liliaceae) 53
Premna corymbosa Rottl. (Verbenaceae) 53
Crataeva nurvala Buch-Ham. (Capparaceae) 54

3. Terpenoids **55**
3.1 Monoterpenes 55
Normal Monoterpenes 56
Iridoids (Cyclopentanoid monoterpenes) 57
Tropolones 57
3.2 Sesquiterpenes 57
3.3 Volatile Oils 59
3.4 Aromatherapy 59
Angelica sinensis Diels. (Apiaceae) 60
Melissa officinalis Linn. (Lamiaceae) 61
Hyssopus officinalis Linn. (Lamiaceae) 61
Tanacetum vulgare Linn. (Asteraceae) 62
Acorus calamus Linn. (Araceae) 62
Alpinia galanga Willd. (Zingiberaceae) 62
Alpinia officinarum Hance 63
Amomum subulatum Roxb. (Zingiberaceae) 63
Anethum graveolens Linn. (Apiaceae) 63
Anethum sowa Roxb. 64
Angelica archangelica Linn. (Apiaceae) 65
Apium graveolens Linn. (Apiaceae) 65
Calendula officinalis Linn. (Asteraceae) 66
Syzygium aromaticum Merr. & Perry (Myrtaceae) 67
Marrubium vulgare Linn. (Lamiaceae) 67
Lavandula officinalis Choix (Lamiaceae) 68
Rosemarinus officinalis Linn. (Lamiaceae) 68
Mentha piperita Linn. (Lamiaceae) 68
Mentha spicata Linn. (Lamiaceae) 69

Thymus vulgaris Linn. (Lamiaceae) 69
Gaultheria procumbens Linn. (Ericaceae) 69
Caesulia axillaris Roxb. (Asteraceae) 70
Carum carvi Linn. (Apiaceae) 70
Chenopodium ambrosioides Linn. (Chenopodiaceae) 71
Chenopodium anthelminticum Linn. 71
Cinnamomum aromaticum Nees. (Lauraceae) 71
Cinnamomum camphora Presl. 72
Cinnamomum tamala Nees & Eberm. 72
Coleus ambonicus Lotur. (Lamiaceae) 73
Coriandrum sativum Linn. (Apiaceae) 73
Cymbopogon citratus Stapf. (Poaceae) 74
Cymbopogon martinii Wats. 74
Eucalyptus globulus Labill. (Myrtaceae) 74
Foeniculum vulgare Mill. (Apiaceae) 74
Myristica fragrans Houtt. (Myristicaceae) 75
Ocimum tenuiflorum Linn (Lamiaceae) 76
Curcuma zedoaria Roscoe (Zingiberaceae) 76
Piper betel Linn. (Piperaceae) 76
Ruta graveolens Linn (Rutaceae) 77
Hyptis suaveolens Poit. (Lamiaceae) 77
Iridoids 78
Barleria prionitis Linn. (Acanthaceae) 78
Barleria cristata Linn. 78
Barleria strigosa Willd. 78
Vitex negundo Linn. (Verbenaceae) 79
Picrorhiza kurroa Linn. (Scrophulariaceae) 79
Gentiana lutea Linn. (Gentianaceae) 79
Sesquiterpenes 80
Chrysanthemum parthenium Pers. (Asteraceae) 80
Nardostachys jatamansi DC. (Valerianaceae) 80
Anthemis nobilis Linn. (Asteraceae) 81
Ageratum conyzoides Linn. (Asteraceae) 81
Artemisia cina Berg. (Asteraceae) 82
Artemisia absinthium Linn 82
Artemisia maritima Linn. 83
Achillea millefolium Linn (Asteraceae) 83
Santalum album Linn. (Santalaceae) 83
Cichorium intybus Linn. (Asteraceae) 84
Pogostemon cablin Benth. (Lamiaceae) 84
Saussurea lappa C.B.Clarke (Asteraceae) 85
Anamirta cocculus W. & A. (Menispermaceae) 85
Elephantopus scaber Linn. (Asteraceae) 85
Inula racemosa Hook.f. (Asteraceae) 86
Inula helenium Linn. 86

Cyperus rotundus Linn. (Cyperaceae) 86
Glucosinolates and Sulphides 87
Lepidium sativum Linn. (Brassicaceae) 87
Tropaeolum majus Linn. (Tropaeolaceae) 87
Armoracia rusticana Gaertn. (Brassicaceae) 88
Capsella bursa-pastoris Medic. (Brassicaceae) 88
Moringa oleifera Lam. (Moringaceae) 89
Allium sativum Linn. (Liliaceae) 89
Allium cepa Linn. 90
Allium porrum Linn. 90
Allium schoenoprasum Linn. 90
Ferula asafoetida Bios. (Apiaceae) 91
3.5 Diterpenes 91
Ginkgo biloba Linn. (Ginkgoaceae) 92
Taxus brevifolia Nutt. (Taxaceae) 93
Swertia chirata Buch.-Ham. (Gentianaceae) 93
Andrographis paniculata Wall. (Acanthaceae) 94
Aleurites fordii Hemsl. (Euphorbiaceae) 94
Baliospermum solanifolium Suresh (Euphorbiaceae) 95
Jatropha gossypifolia Linn. (Euphorbiaceae) 95
Jatropha glandulifera Roxb. 96
Boswellia serrata Roxb. (Burseraceae) 96
Caesalpinia bonduc Roxb. (Caesalpiniaceae) 97
Caesalpinia crista Linn. 97
Calophyllum inophyllum Linn. (Clusiaceae) 98
3.6 Triterpenoids 99
Steroids 99
Sterolins and Saponins 99
3.6a Triterpenes and Steroids 100
Allamanda cathartica Linn. (Apocynaceae) 100
Achyranthes aspera Linn (Amaranthaceae) 100
Trianthema portulacastrum Linn. (Aizoaceae) 101
Vernonia cinerea Less. (Asteraceae) 101
Gossypium arboreum Linn. (Malvaceae) 102
Gossypium herbaceum Linn. 102
Helicteres isora Linn. (Sterculiaceae) 103
Hemidesmus indicus R.Br. (Asclepiadaceae) 103
Ichnocarpus frutescens R.Br. (Apocynaceae) 104
Gymnema sylvestre R.Br. (Asclepiadaceae) 105
Holoptelia integrifolia Planch. (Ulmaceae) 104
Commiphora spp. (Burseraceae) 105
Commiphora wightii Arnott. (Burseraceae) 105
Aerva lanata Juss. (Amaranthaceae) 106
Drypetes roxburghii Hurus (Euphorbiaceae) 106
Chamaesyce hirta Millsp. (Euphorbiaceae) 107
Euphorbia thymifolia Linn. 107

Euphorbia nivulia Buch. - Ham. (Euphorbiaceae) 108
Clerodendrum indicum Kuntze (Verbenaceae) 108
Ficus benghalensis Linn. (Moraceae) 108
Ficus carica Linn. 109
Ficus racemosa Linn. 109
Ficus religiosa Linn. 110
Cissus quadrangula Linn. (Vitaceae) 110
Withania somnifera Dunal (Solanaceae) 110
Citrullus colocynthis Schard. (Cucurbitaceae) 111
3.6b Limonoids and Quassinoids 111
Picrasma excelsa Planch. (Simaroubaceae) 112
Ailanthus excelsa Roxb. (Simaroubaceae) 112
Azadirachta indica A. Juss. (Meliaceae) 113
Brucea javanica Merril (Simaroubaceae) 114
3.6c Saponins 114
Polygala senega Linn. (Polygalaceae) 115
Smilax spp. (Liliaceae) 115
Panax spp. (Araliaceae) 115
Eleutherococcus senticosus Maxim. (Araliaceae) 116
Bupleurum falcatum Linn. (Apiaceae) 117
Platycodon grandiflorum DC (Campanulaceae) 117
Polygala tenuifolia Willd. (Polygalaceae) 117
Akebia quinata DC (Lardizabalaceae) 117
Ziziphus jujuba var. *spinosa* Hu (Rhamnaceae) 118
Abrus precatorius Linn. (Fabaceae) 118
Agave americana Linn. (Agavaceae) 119
Asparagus racemosus Willd (Liliaceae) 119
Chlorophytum borivilianum Sant. & Fernandes (Liliaceae) 120
Fagonia cretica Linn. (Zygophyllaceae) 120
Aesculus hippocastanum Linn. (Hippocastanaceae) 121
Trigonella foenum-graecum Linn. (Fabaceae) 121
Glycyrrhiza glabra Linn. (Fabaceae) 122
Balanites roxburghii Planch. (Simaroubaceae) 122
Bacopa monnieri Wettst. (Scrophulariaceae) 123
Serenoa repens Small (Arecaceae) 123
Taraxacum officinale Weber (Asteraceae) 124
Guaiacum officinale Linn. (Zygophyllaceae) 124
Catunaregam spinosa Tirveng. (Rubiaceae) 125
Centella asiatica Urban. (Apiaceae) 125
Costus speciosus Smith. (Zingiberaceae) 126
Curculigo orchioides Gaertn. (Amaryllidaceae) 126
Tribulus terrestris Linn. (Zygophyllaceae) 127
Actaea racemosa Linn. (Ranunculaceae) 127
3.6d Cardiac glycosides 127
Digitalis purpurea Linn. (Scrophulariaceae) 128
Convallaria majalis Linn. (Liliaceae) 129

Strophanthus kombe Oliv., *S. hispidus* DC (Apocynaceae) 129
Asclepias curassavica Linn. (Asclepiadaceae) 130
Carissa carandas Linn. (Apocynaceae) 130
Cerbera manghas Linn. (Apocynaceae) 131
Calotropis gigantea Ait.f. (Asclepiadaceae) 131
Calotropis procera Ait.f. 132
Corchorus olitorius Linn. (Tiliaceae) 133
Pergularia daemia Choiv. (Asclepiadaceae) 133
Nerium oleander Linn. (Apocynaceae) 134
Leptadenia reticulata W. & A. (Asclepiadaceae) 134
Thevetia peruviana Merril (Apocynaceae) 135
Helleborus niger Linn. (Ranunulaceae) 135
Drimia indica Jessop (Liliaceae) 135

3.7 Carotenoids 136
Lycopersicon esculentum Mill. (Solanaceae) 137
Carica papaya Linn. (Caricaceae) 137
Crocus sativus Linn. (Iridaceae) 138
Nyctanthes arbor-tristis Linn. (Oleaceae) 138
Bixa orellana Linn. (Bixaceae) 138

4. Phenolics **140**

4.1 Simple Phenols 140
4.2 Phenolic Acids 141
4.3 Acetophenones 141
4.4 Phenyl Propanes 141
4.5 Benzophenones 142
Dryopteris filix-mas Schott (Polypodiaceae) 142
Semecarpus anacardium Linn. (Anacardiaceae) 143
Humulus lupulus Linn (Cannabinaceae) 143
Capsicum annum Linn. (Solanaceae) 143
Curcuma longa Linn (Zingiberaceae) 144
Piper methysticum Forst. (Piperaceae) 145
Styrax spp. (Styraceae) 145

4.6 Xanthones 146
4.7 Coumarins 146
Aegle marmalos Correa (Rutaceae) 147
Feronia limonia Swingle (Rutaceae) 148
Psoralea corylifolia Linn. (Fabaceae) 148
Eclipta alba Hassk. (Asteraceae) 148

4.8 Chromones 149
Bergenia ciliata Sternb. (Saxifragaceae) 149
Ammi visnaga Lam. (Apiaceae) 150

4.9 Stilbenes 150
4.10 Lignans 150
Podophyllum peltatum Linn. (Berberidaceae) 151
Podophyllum emodi Wall. 151

Phyllanthus amarus Schum. & Thonn. (Euphorbiaceae) 152
Gmelina arborea Roxb. (Verbenaceae) 152
Piper cubeba Linn. (Piperaceae) 152
Sesamum indicum Linn. (Pedaliaceae) 153
Linum usitatissimum Linn. (Linaceae) 153
Schisandra chinensis Linn. (Schisandraceae) 154
Aleurites moluccana Willd. (Euphorbiaceae) 154
4.11 Flavonoids 155
4.11a Anthocyanins 155
4.11b Aurones 156
4.11c Biflavonyls 156
4.11d Catechins, Flavan-3, 4-diols and Proanthocyanidins 156
4.11e Chalcones 157
4.11f Flavones and Flavonols 157
4.11g Flavonones and Flavononols (Dihydroflavones and Dihydro Flavonols) 158
4.11h Isoflavones, Isoflavonones and Homoisoflavones 158
4.11i Neoflavanoids 159
Vaccinium myrtillus Linn. (Vacciniaceae/Ericaceae) 159
Crataegus laevigata DC & C. *monogyna* Linn. (Rosaceae) 160
Sambucus nigra Linn. (Caprifoliaceae) 160
Hibiscus rosa-sinensis Linn. (Malvaceae) 160
Ixora coccinea Linn.(Rubiaceae) 161
Fagopyrum esculentum Moench. (Polygonaceae) 161
Camellia sinensis Kuntze (Theaceae) 162
Oroxylum indicum Vent. (Bignoniaceae) 163
Biophytum sensitivum DC. (Oxalidaceae) 163
Woodfordia floribunda Salisb. (Lythraceae) 164
Gardenia gummifera Linn. f. (Rubiaceae) 164
Gardenia resinifera Roth. 165
Bauhinia purpurea Linn. (Caesalpiniaceae) 165
Bauhinia racemosa Lam. 165
Bauhinia tomentosa Linn. 166
Bauhinia variegata Linn. 166
Cardiospermum halicacabum Linn. (Sapindaceae) 166
Canscora decussata Roem. & Schult. (Gentianaceae) 167
Glycine max Merril (Fabaceae) 167
Iris germanica Linn. (Iridaceae) 168
Pterocarpus marsupium Roxb. (Fabaceae) 168
Tephrosia purpurea Pers. (Fabaceae) 169
Derris indica Bennet (Fabaceae) 169
Mallotus philippensis Don. (Euphorbiaceae) 170
Silybium marianum Gaertn. (Asteraceae) 170
4.12 Quinones 171
4.12a Benzoquinones 171

4.12b Naphthaquinones 171
4.12c Anthraquinones 171
Embelia ribes Burm.f. (Myrsinaceae) 172
Arnebia nobilis Rachinger (Boraginaceae) 172
Stereospermum suaveolens DC (Bignoniaceae) 173
Plumbago indica Linn. (Plumbaginaceae) 173
Aloe barbadensis Mill. (Liliaceae) 173
Cassia Linn. (Caesalpiniaceae) 174
Cassia occidentalis Linn. (Caesalpiniaceae) 174
Cassia tora Linn. 175
Cassia senna Linn. 175
Cassia fistula Linn. 176
Cassia angustifolia Vahl. (Caesalpiniaceae) 177
Rheum palmatum Linn. (Polygonaceae) 177
Rhamnus purshiana DC. (Rhamnaceae) 178
Rhamnus alnus Mill. (Rhamnaceae) 178
Andira araroba Aguiar (Fabaceae) 179
Rubia cordifolia Linn. (Rubiaceae) 179
Pterocarpus santalinus Linn. (Fabaceae) 179
Hypericum perforatum Linn. (Hypericaceae) 180
4.13. Tannins 180
Barringtonia acutangula Gaertn (Lecythidaceae) 182
Hamamelis virginiana Linn. (Hamamelidaceae) 182
Syzygium cumini Skeels (Myrtaceae) 183
Saraca asoca De Willde (Caesalpiniaceae) 183
Thespesia populnea Sol. (Malvaceae) 184
Terminalia chebula Retz. (Combretaceae) 184
Terminalia bellirica Roxb. (Combretaceae) 185
Terminalia arjuna W.&A 185
Symplocos cochinchinensis Moore (Symplocaceae) 185
Adenanthera pavonina Linn. (Mimosaceae) 186
Acacia catechu Willd. (Mimosaceae) 186
Acacia nilotica subsp. *indica* Brenan (Mimosaseae) 187
Albizzia lebbeck Willd. (Mimosaceae) 187
Albizzia procera Benth. 188
Albizzia odoratissima Benth. 188
Butea monsperma Taub. (Fabaceae) 188
4.14 Antioxidant Therapy 189

5. Gums and Mucilages **192**
Echinacea angustifolia DC. (Asteraceae) 192
Plantago ovata Forsk. (Plantaginaceae) 193
Hygrophila auriculata Heine (Acanthaceae) 193
Sterculia spp.—(Sterculiaceae) 194
Anogeissus latifolia Wall. (Combretaceae) 194

Abelmoschus esculentus Moench (Malvaceae) 194
Abutilon indicum Sweet. (Malvaceae) 195
Abutilon hirtum Sweet. 196
Acacia senegal Willd. (Mimoseae) 196
Ulmus rubra Muhl. (Ulmaceae) 196
Alcea rosea Linn. (Malvaceae) 197
Althea officinalis Linn. (Malvaceae) 197
Bambusa arundinacea Roxb. (Poaceae) 198
Tabashir 198
Basella alba Linn. (Basellaceae) 198
Benincasa hispida Cogn. (Cucurbitaceae) 199
Amorphophallus campanulatus Blume (Araceae) 199
Ceratonia siliqua Linn. (Caesalpiniaceae) 200
Macrocystis spp. (Laminariaceae) 200
Gelidium cartilagineum Gaillon (Rhodophyceae) 201
Chondrus spp. and *Gigartina* spp. (Rhodophyceae) 201

PART B: PRIMARY METABOLITES

6. The Primary Metabolics **205**
6.1 Carbohydrates 205
6.2 Aminoacids and Proteins 206
Citrus spp. (Rutaceae) 208
Emblica officinalis Gaertn. (Euphorbiaceae) 209
Portulaca oleracea Linn. (Portulacaceae) 209
Tamarindus indica Linn. (Caesalpiniaceae) 210
Oxalis corniculata Linn. (Oxalidaceae) 210
Garcinia indica Choiss. (Clusiaceae) 210
Garcinia mangostana Linn 211
Garcinia combogia Desv. 211
Garcinia morella Desv. 211
6.3 Fatty Acids, Triacylglycerols and Glycolipids 212
Oenothera biennis Linn. (Onagraceae) 212
Hydnocarpus wightiana Bl. (Flacourtiaceae) 213
Ricinus communis Linn. (Euphorbiaceae) 213
Exogonium purga Benth. (Convolvulaceae) 213
Ipomoea orizabensis Led. (Convolvulaceae) 214
Operculina turpethum Silva Manso (Convolvulaceae) 214

General References 215

References 216

Appendix 224

Index 230

1

Introduction

About eighty per cent of the world depends on herbal-based alternative systems of medicine. Except for homeopathy, the activities of these curative plants are evaluated by their chemical components. An estimated 70,000 plants (including the lower plants) are used in medicine. Indian Ayurveda utilizes about 2000 plants to cure different ailments. The Chinese system depends on the 5757 plants listed in the *Encyclopedia of Traditional Chinese Medicinal Substances*. Japanese and Korean systems of medicine also include a large number of medicinal herbs. In WTO perspective, all these plants are our common heritage. Utilizing all these plants for human welfare has mooted the concept of the herbal medicine or phytotherapy. Herbal medicine is now expanding at an astonishing pace due to the great inputs from ethnomedicinal practices being pooled from all over the world.

Every useful plant owes its activity/utility to a number of ingredients whether it is food, clothing or shelter. Similarly, every medicinal herb contains a number of active constituents facilitating its manifold curative activities. A few of these compounds such as reserpine, taxol, vincristine have been isolated/synthesized in a large scale and are used singly in allopathic systems.

When one views the plant as a capable organism synthesizing its food and other chemicals needed for its maintenance, which includes protection from microbes, animals, stress and healing of wounds and infection, he may be mystified by the large number of chemicals these plants are applying for these different uses. Since external medication is not possible for the plant it has to do all these functions alone and it does it wonderfully. All these plethora of compounds with myriads of activities can be roped into animal/human systems for the same properties. This belief should be the logic of herbal medicine/culture.

A single metabolome, the sum total of primary and secondary metabolites of a plant, contains about 4,000 compounds (It may even go up to 10,000). The total number of compounds present in the plant kingdom is about 2,00,000. All these compounds are used by plants for various purposes such as nutrition, maintenance, reproduction, healing, defense, offense, etc. We generally use plants only as nutrient sources and, to some extent, for healing processes. When we consume a plant/product as food, we receive not only nutrients from them but also a large variety of "other compounds" the function of which in our system is relatively unknown. Many of the main food plants (cereals and pulses) contain a fairly good amount of polyphenolic antioxidants (Latha & Daniel, 2001). Vegetables, fruits, nuts, etc. are also good sources of secondary metabolites like polyphenols, carotenoids, volatile oils, organic acids, waxes, etc. (Daniel, 1989). The role of these compounds, which are often taken unintentionally in our body, is not assessed. All these compounds play a role in protecting and maintaining our system. The great role of "dietary fibre" (earlier known by a contemptuous term 'roughage') in reducing cholesterol and sugar levels in the body, and other digestive problems has been realized only recently.

The large conglomerate of organic compounds present in a herbal extract, used for medication, casts a great doubt on the so-called "active component". This extract may contain a good number of oligosaccharides, polysaccharides, starches (if a tuber is used), sugar derivatives, aminoacids (both protein and non-protein, including essential amino acids), peptides, proteins, fatty acids (including essential fatty acids), fatty alcohols, vitamins, and minerals besides the "privileged" secondary metabolites. It is these very same compounds, which make and regulate the body, on consuming the plants as foods. Therefore, these additional primary metabolites may have a very positive role when they are taken as a medicinal extract. Earlier, every phytochemist was looking for active components like alkaloids, saponins and tannins in plants. Terpenoids and phenolics, especially the latter, received very scanty attention.

Many a times, the group phenolics are cursorily mentioned in the account of a medicinal plant. But the renewed interest in antioxidants, after the realization of their active role in diseases such as atherosclerosis, stroke, cerebral thrombosis, diabetes, Alzheimer's and Parkinson's diseases and cancer (Tiwari, 2004) have firmly put these compounds on a pedestal as pharmacologically active components. Since phenolics are very effective antioxidants, besides their curative role, they play varied roles in: (1) protecting other active components in a medicinal preparation; (2) protect the membranes and tissues from oxidation at the site of medicinal action; or (3) protect other antioxidants (needed elsewhere) from oxidation. The contribution of phenolics as antimicrobials in providing a microbe-free environment, for the other drugs to act, is also significant. But for phenolics, no other group of compounds would have any role in maintaining the system from wear and tear. Alkaloids and terpenoids have mostly curative actions on the system. The antimicrobial terpenes (mono- and sesquiterpenes) also, if present in an extract, significantly support the action of other drugs.

I do have my doubts on the authenticity of some of the compounds reported from certain medicinal plants. Most of the reports published before 1980 should be treated with caution, not because of doubts on the authenticity of the identification of the compounds, but because of the incorrect identity of the plants from which the compounds are extracted. It the source plant is wrong or adulterated, then the reports are highly unreliable. Only recently, scientific journals insist on keeping herbaria of all the plants screened, before publishing the results on these plants. A fresh screening of all the plants for all the constituents is the need of the hour. Seasonal/geographical/

ecological variations of the phytochemicals of drug plants is another serious area of concern.

Ayurveda and Siddha, the predominant herbal practices in India, are considered sciences of life, prevention of ailments and longevity. They are not a mere collection of healing practices, but contain methods of maintaining and prolonging the health rather than just simple healing. The **holistic approach** of these practices, which advocates the involvement of both mind and body (and which complement each other) in maintaining the body and healing the same, is **a way of living itself**. This is in sharp contrast with the allopathic system of medicine, in which the diseases are identified first and then the healing therapies are applied. When one system or the organ of the body is infected/affected, all other systems of the body are similarly affected. For example, an infection in the stomach weakens the circulatory system, urino-genital system, respiratory system and even affects the psyche of the patient. The body becomes weak; immunity goes down and this paves way for secondary infections or problems. But normally a remedy in the form of a laxative is meant only for the alimentary canal and does not take care of the other systems affected. Therefore, even when a patient is cured (of the stomach problem), he remains weak. In contrast, when a herbal formulation (in Ayurveda, a number of plants are involved in a formulation) is given, the multitude of components, including a large amount of primary metabolites, are available to the body to repair all the systems. The body is a wonderful machine and it repairs itself, using all these components. And that is the reason why a person becomes healthy and remains so after a herbal (say Ayurvedic) treatment. To quote an analogy, herbal therapy is like a farmer enriching the field to get a better crop, whereas the other systems are like enriching the seed while neglecting the field.

A knowledge of chemical components of a plant is essential for quality control analyses of the plant, extract or any formulation containing them. A compound or a group of compounds present can serve as a "biomarker" and the presence and concentration of the same can be followed to decide on the genuineness of the drug /formulation. Any component other than the biomarkers present indicates adulteration. Not many plants are studied for these markers. A knowledge of these compounds and their specific analytical methods will facilitate the herbal industry in checking adulteration and thus raise its standards.

The data on the pharmacological action of some of the day-to-day compounds which we come across are also interesting. Duke (1997) describes ferulic acid, gentisic acid, kaempferol glycosides and salicylic acid as pain relievers while ascorbic acid, cinnamic acid, coumarin, myricetin, quercetin and resveratrol are explained to be anti-inflammatory. Even the variety of chemicals and their richness (concentration) in a medicinal herb is of great value in assessing its property. Duke's data base states that both coriander and liquorice contain 20 chemicals with antibacterial action; oregano and rosemary have 19; ginger 17; nutmeg 15; cinnamon and cumin 11; black pepper 14; bay 10 and garlic 13. Quantity wise, liquorice contains up to 33% bactericidal compounds (dry weight basis), thyme 21%, oregano 8.8%, rosemary 4.8%, coriander 2.2% and fennel 1.5%. Needless to say, the above data explain the drug action of the said plants.

Since this book deals with the chemical components of medicinal herbs, I have divided the chapters based on the classes of compounds. The major classification of plant chemicals to primary and secondary metabolites is followed. But I experienced problems in placing gums and mucilages in the proper place. Though they are technically primary metabolites, they are stored in plants, never metabolized and are of survival value as other secondary metabolites. I, therefore, prefer to keep them in secondary metabolites. Secondary metabolites are placed first due to their drug action and alkaloids, terpenoids and phenolics have been further subdivided.

PART A
Secondary Metabolites

Secondary Metabolites

Secondary metabolites, which do not get involved in the primary metabolism of plants, are a group of compounds whose role has always been underestimated. The great misconceptions on the function of these compounds lead to naming them by erroneous terms like "waste products", "end products", "excretory products", etc. But now there is a genuine realization of the varied roles of these compounds, most of which are as adaptational and survival mechanisms. Thus, alkaloids are defensive agents, deterring the herbivores due to their bitter taste; volatile oils act as pheromones (for attracting insects for pollination) and protect the plant from microbes (and sometimes suppresses neighbouring plants, as in case of *Eucalyptus*) and diterpenes or triterpenes, when present in resin, perform wound healing and antimicrobial functions. Limonoids and quassinoids are insect repellants and thus protect the plant (especially wood) from insect attack. Phenolics are the best antioxidants, protecting cellular membranes and tissues containing lipids from oxidation. Among them, anthocyanins and flavonoids are flower pigments, which make the flowers visible to the "color blind" insects, and thus act as pollinator guides. All sulphur-containing compounds are antimicrobial in nature.

The manifold functions of these compounds in plants reiterate that they are not aberrations in biosynthesis in plants and that their syntheses are genetically controlled mechanisms. They are never formed in any of the primary metabolic pathways but are prepared by **specially designed pathways, the secondary metabolic pathways.** All the molecules formed by these specialized secondary metabolic pathways are designated as secondary metabolites. The adaptive significance of each one of these compounds is being understood now. It is now confidently proclaimed that the plants survive and evolve due to their secondary metabolites. Each group of plant specializes in one or the other group of compounds which enables them to survive better, be it the pollination of flowers, dispersal of seeds, warding of pathogens and herbivores, or removing competition by suppressing the neighbouring plants. It is these compounds that make the plants strong in the face of adversities, enemies or competition. The plants, specialized in compounds, performed better, evolved further and became the large taxa.

It is generally believed that secondary metabolites, once formed, are never metabolized. But studies on the dynamics of secondary product metabolism indicate that there is a definite **turnover** of these compounds evidenced by diurnal variation, seasonal variation and disappearance at various stages of development (as in the case of bitter tannins which are lost when certain fruits ripen and disappearance of anthocyanins when the leaves mature). Pulse labelling experiments have proved that the concentrations of isoflavones (*Phaseolus aureus*), alkaloids (*Atropa* and *Papaver*) and flavonols and cinnamoyl derivatives of many plants remain at constant levels as a result of simultaneous and equally rapid synthesis and subsequent consumption.

When we ingest a herbal medicine, we are adapting the very same compounds which the plants use themselves. Thus, the antioxidants of the plant can be used by man for the same function of protecting his own system from free radicals and oxidative damage leading to ageing. The same carotenoids and ascorbic acid, which protect chlorophyll and other molecules, can be used to protect human molecules. Cardiac glycosides, by which a plant kills a herbivore by stopping its heartbeat, can be used as cardiotonics. The antimicrobial volatile oils can be used to repel the attack of a microbe in our system. Man, for his welfare, can use these properties and the other metabolic changes caused by some of the compounds in our body profitably. In addition, plants contain a number of molecules, which mimic our molecules as in the case of isoflavonoids that simulate estrogens. They can be employed in our system to rectify a derailed pathway.

There are **three** major classes of secondary metabolites. The largest group is **alkaloids**, followed by **terpenoids** and **phenolics**. Sulphur-containing compounds, though biogenetically unrelated to terpenoids, are included in the sections of volatile oils. Nitro compounds such as hydroxylamine, indigo, aristolochic acid and betacyanins are kept in the chapter on alkaloids. **Gums and mucilages** are a group of polysaccharides included in secondary metabolites because of their functions.

2

Alkaloids

Alkaloids comprise the largest class of secondary plant substances, at present numbering more than 7,000. The term 'alkaloid' means 'alkali-like substance'. A typical alkaloid is a basic plant product possessing a nitrogen-containing heterocyclic ring system and exhibiting marked pharmacological activity.

Alkaloids form a very heterogeneous group. To define them as just basic, organic nitrogenous plant metabolites seems most indiscriminate, but alkaloids exhibit no basic unity in either their chemical structures, or their biosyntheses. Besides, they also elicit a wide range of physiological responses.

The alkaline nature of alkaloids is due to nitrogen. Alkaloids may contain a single N atom, as in atropine; or as many as 5, as in ergotamine. The N may occur as a primary amine, i.e. RNH_2, as in mescaline; a secondary amine, i.e. R_2NH, as in cytosine; as a tertiary amine, i.e. R_3N, as in physostigmine; or as a quaternary ammonium ion, i.e. $R_4N^+X^-$, as in tubocurarine chloride. Except in quaternary ammonium compounds, the N atom possesses a pair of unshared electrons, which makes the alkaloids alkaline. The degree of basicity varies from slight to moderate to strong, depending on the position of the N in the skeleton, and on that of other functional groups. In the cell sap, alkaloids exist as cations of the salts of various organic acids.

Apart from C, H and N, most alkaloids also contain O, also, and are therefore, crystalline solids. The few alkaloids like coniine and nicotine, which lack O, are liquids at room temperature. Almost all are colourless; exceptions being berberine and serpentine, which are yellow and sanguinarine, which is brownish red. Alkaloids are usually insoluble in water, or only sparingly so, but are freely soluble in ether, $CHCl_3$ and other relatively non-polar solvents. Their salts, formed by their treatment with acids, are soluble in water but insoluble in non-polar solvents. This property is made use of in the extraction, purification and estimation of alkaloids. Alkaloidal salts are crystalline, often with characteristic crystal forms and habit. Alkaloids can be precipitated out of their aqueous or acid solutions by a number of substances like picric acid and tannic acid.

Most alkaloids are active optically, and are usually *l*-rotatory. Normally, only one of the isomers occurs naturally in a plant, and only occasionally do racemic mixtures occur. In most cases, the isomers differ in their physiological activities. Almost all alkaloids absorb UV light, and possess characteristic absorption spectra. This property is made use of in characterizing the group.

All alkaloids occur in plants, and a few of them in animals too. Almost all of them can be synthesized chemically. Present studies indicate a rather restricted dispersion of alkaloids amongst plants. Among Cryptogams, they occur in certain fungi like *Claviceps* and *Amanita* and also some ferns. Only the Taxaceae and Gnetaceae, amongst the Gymnosperms, show alkaloids. Of Angiosperms, monocots generally do not produce alkaloids, but the Amaryllidaceae and Liliaceae, Stemonaceae, Dioscoreaceae, Arecaceae, Poaceae and Orchidaceae are exceptions. Among dicots, the Fabaceae, Papaveraceae, Ranunculaceae, Rubiaceae, Solanaceae, Berberidaceae, Apocynaceae, Asclepiadaceae, Asteraceae, Menispermaceae and Loganiaceae include most of the alkaloid-yielding plants. Alkaloids are entirely absent from some plant families. Usually alkaloids with complex structures are characteristic of specific plant families, e.g. colchicine in the Liliaceae. Treibs (1953) suggests that some interrelationships exist between alkaloids and volatile terpenes in the sense that they tend to be mutually exclusive.

Alkaloids, as a rule, do not occur singly in plants. The plant usually produces a series of alkaloids, which may differ only slightly in physical and chemical characters. In the plants, alkaloids may be systemic, i.e. distributed throughout, or restricted to specific organs like roots (Aconite, Belladona), rhizomes and roots (Ipecac, *Hydrastis*), stem barks (*Cinchona*, Pomegranate), leaves (*Hyoscyamus*, Belladona), fruits (Pepper, *Conium*) or seeds (Nuxvomica, *Areca*).

From a chemotaxomical viewpoint, Hagnauer (1966) placed the alkaloids into the following three classes:

1. True Alkaloids
   These have nitrogen containing heterocyclic ring skeleton derived from a biogenetic amine formed by decarboxylation of an amino acid.
2. Protoalkaloids
   These are derived from amino acids or biogenetic amines—their methylated derivatives, but lack heterocyclic nitrogen. The nitrogen may be present in an aliphatic side chain, as in colchicine.
3. Pseudoalkaloids
   These basic compounds do contain nitrogen, but the carbon skeleton is derived from mono-, di- or tri-terpenes, sterols, or acetate-derived polymers.

Classification of alkaloids is usually based on their chemical structures. The structural complexity ranges from monocyclic (e.g. coniine) to heptacyclic (e.g. kopsine). The most common skeletons found in alkaloids are indole, isoquinoline, quinoline, pyridine, piperidine, tropane, purine, pyrrole, pyrrolidine and steroid.

The probable amino acid precursors of alkaloids are:

(a) Ornithine, in the biosynthesis of tropane and pyrrolidine alkaloids;
(b) Lysine and cadavarine, in the biosynthesis of piperidine and lupine alkaloids;
(c) Phenylalanine and tyrosine, in the biosynthesis of isoquinoline alkaloids and protoalkaloids; and
(d) Tryptophan, in the biosynthesis of indole and quinoline alkaloids.

The role of alkaloids in plants remains a mystery. Possibly, they may serve one or more of the following functions (Robinson, 1975).

(a) They are non-toxic excretory products of toxic compounds.
(b) They serve as nitrogen reserve food materials.
(c) They are protective; being poisonous to parasites and browsers.
(d) They act as growth regulators. Lupine alkaloids inhibit germination, some alkaloids remove the inhibitory effects of tannins, and a few act as growth stimulators.
(e) Being basic, they help the plant maintain its ionic balance by replacing mineral bases.

The pharmacological actions of alkaloids are:

1. Analgesics and narcotics, e.g. morphine and codeine
2. Emetics, e.g. emetine
3. Central stimulants, e.g. strychnine, brucine
4. Local anaesthetics, e.g. cocaine
5. Myotics, e.g. physostigmine, pilocarpine
6. Antihaemorrhagic, e.g. hydrastine
7. Antispasmodics, e.g. atropine, hyoscyamine
8. Vermifuges, e.g. pelletierine
9. Aphrodisiacs, e.g. yohimbine
10. Anti-hypotensives, e.g. ephedrine
11. Anti-hypertensives, e.g. reserpine
12. Cardiac repressants, e.g. quinine
13. Diaphoretics, e.g. pilocarpine
14. Muscle paralyzers, e.g. tubocurarine
15. Antitumour agents, e.g. camptothecine

However, most alkaloids are toxic at higher concentrations, and the physiological response they elicit depends chiefly on the dosage.

2.1 Alkaloidal amines

Possessing an aliphatic N-atom, these compounds are often designated 'protoalkaloids'. Of the fifty and odd members included in this group, a majority occur in two families, Chenopodiaceae and Cactaceae, and a few like ephedrine are widely distributed in the plant kingdom.

Ephedra sinica Stapf. (Ephedraceae)

Ma-Huang, Ephedra

This monoecious (rarely dioecious) shrub is 30-60 cm in height and has a woody stem creeping in the soil and many erect aerial stems. The leaves are sheath-like and membranous; the basal ones half connate and the upper ones half divided with a triangular 2-nerved free portion. Inflorescences are cone-like. The male cones are broad and ovoid, terminal or axillary in clusters of 3-5. The bracts of the solitary terminal female cone increase in size, become fleshy and give the cone the appearance of a red berry.

The slender stem, which forms the drug, contains 0.5-2.5% alkaloids. Ephedrine and pseudoephedrine, the principal alkaloids, amount to 90% of alkaloids. Other alkaloids are methylephedrine, methylpseudoephedrine, norephedrine, norpseudoephedrine, ψ-ephedrine, *l*-nor-ephedrine, *d*-nor-ψ-ephedrine, N-methylephedrine, N-methyl-ψ-ephedrine, ephedrine benzylmethylamine, 2,3,4,trimethyl-5-yl-phenyloxazolidine, and ephedroxane. Roots contain macrocyclic spermine alkaloids, ephedradine A-D. Bisflavonols, glycoflavones, mahuanin A&B and ephedranin are the other compounds present. The ephedrine content is found to be maximum in winter. From the crude drug of *Ephedra* spp. esp. *E. distachya*, five glycans, ephedrans A-E are isolated. The aerial parts of *E.sinica* yield flavonoids such as tricin, apigenin (both together constituting 50% of flav.), kaempferol, herbacetin and 3-methoxy herbacetin. (Handa et al. 1992; Grue-Sorenson & Spenser, 1993; Liu, et al. 1993)

Ephedrine provides relief in asthma and hay fever. Ephedrine and pseudoephedrine act as cardiac stimulants. Ephedroxane is anti-inflammatory.

Alhagi pseudalhagi Desv. (Fabaceae)

Syn. *A. camelorum* Fisch.
A. maurorum Medic.

Jawasa, Camel thorn

This is a glabrous spiny shrub having small oblong-obtuse leaves, red papilionaceous flowers in axillary racemes forming panicles and falcate pods containing 6-8 sub-reniform seeds. A native of central Asia, this plant is common in the arid zones.

The twigs contain alkaloids such as β-phenylethylamine; its N-methyl derivative, N-methyl tyramine, hordeine, 3,4-dihydroxy-β-phenethyl trimethylammonium hydroxide, its 3-methoxy deriv., N-methyl mescaline and salsodine and traces of betaine and choline. The roots also contain same alkaloids but in lesser amounts. Leaves are rich sources of rutin whereas the roots and fruits contain a good amount of tannins. The whole plant is found to contain tannins, flavonoids, coumarins, ascorbic acid, mucilage, and essential oils. Roots are reported to contain quercetin, gum resin and wax. A neutral proteinase, alhagain, is also reported from the plant.

The plants of this species growing in Turkey, Iraq and Iran produce a sweet sugary excretion in the form of small, round opaque tears known as Alhagi manna. It contains sucrose (42%), melizitose (25%) and other reducing sugars (9%). This is used as an expectorant and laxative. The entire plant is used as a laxative, diuretic and expectorant. Its infusion is said to be diaphoretic.

Catha edulis Forsk (Celastraceae)

Khat, Abyssinian Tea

C. edulis is a small tree with white bark, ovate-lanceolate leaves, small white flowers in cymes, woody capsules splitting into 3 valves and small brown seeds. A native of Africa, it is now grown in gardens.

The leaves yield phenylpropanoid amines such as cathinone as well as *nor*-ψ-and *nor*ephedrines and other alkaloids such as cathine (nor-pseudoephedrine), cathidine A,B,C and D and cathedulins. Total alkaloid fraction of leaves amounts to 20-28%. Other alkaloids reported from leaves and young shoots are cathinine, ephedrine, norephedrine, l-phenyl – 1,2-propane dione and l-cinnamomyl ethylamine. Also present in leaves are kaempferol, quercetin, myricetin, ampelopsin and dulcitol.

Cathinone is a major CNS-active compound and the leaves exhibit sympathomimetic properties. The leaves are chewed or consumed as tea to attain a state of euphoria and stimulation.

Sida cordifolia Linn. (Malvaceae)

Bala

S. cordifolia is a hairy, branched shrub with stellate pubescence, cordate-crenate leaves, yellow flowers (with monadelphous, reniform anthers) and a schizocarpic fruit enclosed by the calyx containing two long awns on each mericarp. Common all over India, this plant grows wild.

Roots, the official drug, contain mucilage, ecdysterone, a variety of alkaloids such as β-phenethylamines, ephedrine and pseudoephedrine; carboxylated tryptamines (S-(+) – N-methyl tryptophan methylester and hypaphorine); quinazolines like vasicinone, vasicine and vascinol and sitoindoside-X, an acylsteryl glycoside (Begerhotta and Banerjee, 1985; Ghosal et al., 1988; Deepak, 1999).

Bala is one of the important general tonics (*rasayana* drugs) and is especially useful in rheumatism. Sitoindoside-X exhibits adaptogenic and immunostimulant properties. It increases anti-salmonella antibodies and is anti-plaque and anti-microbial in nature.

A number of other species of *Sida* used as **bala** are *S. rhombifolia* Linn. (**Mahabala**), *S. acuta* Burm.f., *S. spinosa* Linn. (**Nagabala**-containing glyceryl-1-eicosanoate and 20-hydroxy,24-hydroxymethyl ecdysone) (Darwish and Reinecke, 2003), and *S. cordata* Borss., *Abutilon indicum* Sweet and *Urena lobata* Linn are considered as **Atibala.**

Urtica dioica Linn. (Urticaceae)

Stinging nettle

U. dioica is a hardy dioecious herb armed with stinging hairs all over the plant. The stem is grooved and the leaves are ovate/lanceolate, cordate and serrate. The flowers are green, borne in axillary cymes.

The stinging hairs contain acetyl choline, histamine and 5-hydroxytryptamine. Acetyl choline is also present in leaves, roots and stem. The stem contains histamine, leaves yield betaine and choline. The leaves are rich sources (upto 30%) of proteins, containing all essential amino acids. Seeds yield oil (33%) and are rich in linoleic and oleic acids.

The property of irritation caused by the herb has been used to excite activity in paralyzed limbs, for arthritis and ingested internally for treatment of haemoptysis and haemorrhages (for paralysis and acute arthritis). The treatment is called flagellation/urtication in which the patient is slapped or pricked with a bundle of fresh twigs for 1 or 2 minutes once or twice a day. An infusion of leaves is hypoglycemic, diuretic, expectorant and a blood-purifier and used to control excessive menstrual flow. The tender leaves and shoots are used as vegetables and for preparation of soups (Duke, 1997). The plant loses its irritant property on boiling in water.

2.2 Diterpenoid Alkaloids

Out of the total 90 diterpenoid alkaloids known, a majority (more than 70) are isolated from two Ranunculaceous genera *Aconitum* and *Delphinium*. The structural elucidations of all the alkaloids from these two genera are not in any way complete. But all of them are found to be derived from complex polyhydric alcohols probably based on $C_{19}H_{28}NH$, which posseses a diterpenoid skeleton. The other genera yielding similar alkaloids are: (1) *Inula* (Asteraceae); (2) *Garrya* (Garryaceae); and (3) *Erythrophleum* (Fabaceae). The pharmacologically important alkaloids are only three—aconitine, artisine and ajacine.

Aconitum napellus Linn. (Ranunculaceae)

Ativish/Bacchanag, Aconite

The roots of *A. napellus*, a native of Alps, and other mountainous regions of Europe and Asia, provide the drug Aconite. It is an erect, showy, herbaceous perennial with turnip-like tuberous roots, large-lobed leaves, blue-violet flowers in spike-like terminal racemes and follicular fruits.

Total alkaloid yield of the roots varies from 0.3 to 1.5%. Two types of alkaloids—aconitines, and atisines—have been recognized in the drug. Aconitines are diacyl esters of polyhydric amino alcohols and are highly poisonous. Atisines are simple amino alcohols and possess less toxicity. Aconitine. $C_{34}H_{47}O_{11}N$- is the major alkaloid amounting to 30% of the total alkaloid fraction and yield aconine ($C_{25}H_{41}O_9N$), acetic acid and benzoic acid on hydrolysis. Hypaconitine, neopelline and neoline are the other alkaloids present in appreciable quantities. Also present are trace amounts of ephedrine, sparteine, aconitic acid and a large amount of starch.

Aconite is a highly potent and quick-acting poison having drastic cardiac effects. The drug is now used as a local analgesic in liniments. Since this is a poisonous drug, Ayurvedic physicians keep the roots in cow's urine or subject it for prolonged heating. It is then used as a cure for fever and rheumatism and as an expectorant.

The other species of *Aconitum* used in medicine are:

Aconitum ferox Wall

Bacchanag, Indian aconite

This is a herbaceous perennial with tuberous dark brown ovoid/ellipsoid roots, palmately 5-fid cordate/reniform leaves, pale blue flowers borne on a terminal raceme and oblong follicles enclosing obovoid winged seeds. It is common in alpine regions of Kashmir and Nepal.

The alkaloid content of tubers reaches upto 5%. The major alkaloids are pseudaconitine, chasmaconitine, indaconitine, bikhaconitine, veratryl pseudaconitine, diacetyl pseudaconitine, picro-aconine, aconine, veratryl-ψ-aconine and di-Ac-ψ aconitine. The drug is highly toxic. The tubers are used only after soaking them in milk for 2-3 days to reduce their depressant action on the heart and, instead, become mild cardiotonics. They are then used as diaphoretic, diuretic, anodyne and antidiabetic. The paste of tubers is useful in neuralgia, muscular rheumatism and inflammatory joint affections. It is also used as sedatives and diaphoretics.

Aconitum heterophyllum Wall

This tall herb, common in alpine or subalpine Himalayas, possesses paired white tuberous roots, variable leaves (lower long-petioled, upper amplexicaul), racemes bearing blue flowers and follicles enclosing blackish obpyramidal seeds.

The alkaloid content of roots is upto 0.8%. The principal alkaloids are atisine (50%), heteratisine, heterophyllisine, heterophylline and heterophyllidine. Other alkaloids present are atidine, hetidine, benzyl heteratisine and hetisinone.

Atisine exhibits lesser toxicity than aconitine. The entire aqueous extract showed marked hypotensive activity. Roots act as antifertility agents. Tubers are used against dyspepsia, acute inflammation, loss of memory, hysteria and piles. It is also an efficient febrifuge and a tonic to remove debility after malaria. The tubers are sometimes used as a vegetable.

Aconitum chasmantham Stapf.

Patisa, Indian Napellus

A plant very similar to *A. napellus* but differing in having deeply grooved and wrinkled tubers. The alkaloid content is about 3% and the major components are indaconitine, chasmaconitine, chasmanthinine, chasmanine and homochasmanine. The medicinal uses of this drug are same as *A. ferox*.

Larkspur–*Delphinium ajacis* L. yields a number of poisonous alkaloids like ajacine, ajacinine and ajaconine, which are very similar to aconitine. These alkaloids are often used as parasiticides.

2.3 Imidazole Alkaloids

Imidazole alkaloids possess an imidazole-glyoxaline-nucleus, very similar to that of histidine. Although such a ring occurs in purines, they are treated as separate because of the differences existing in biosynthetic pathways. Imidazoles are a small group of 13 alkaloids reported from the Rutaceae, Euphorbiaceae and Fabaceae, of which pilocarpine is the only economically important alkaloid.

Pilocarpus spp. (Rutaceae)

Jaborandi

The various species of *Pilocarpus.*, i.e. *P. jaborandi* Holmes, *P. microphyllus* Stapf, and *P. pinnatifolius* Lam., indigenous to Brazil, are sources of pilocarpine.

The dry imparipinnate leaves with ovate leaflets yield about 0.7-1.0% alkaloids consisting of pilocarpine (a lactone of pilocarpic acid – 0.5-0.9%) isopilocarpine, pilosine, isopilosine and pilocarpidine and 0.5% vol. oil. The leaves tend to lose the alkaloidal content on storage.

The salts of pilocarpine cause contraction of the pupil of the eye—a property antagonistic to atropine and so finds great use in opthalmic practices. They also stimulate secretion of saliva.

2.4 Indole Alkaloids

With their 600 odd members, indoles form the largest class of alkaloids. They possess an indole or a near indole group and, with a few exceptions, are derived from tryptophan/tryptamine and a 10-carbon monoterpenoid precursor. This monoterpene exists in three different skeletons designed as aspidosperma, corynanthe and iboga and so the alkaloids are classified into aspidosperma, corynanthe and iboga types. As a grouping like this gives only a very faint idea about the complexity and heterogeneity existing within the indoles, many different systems of classification are proposed of which none is too convenient to be reproduced here. Based on chemotaxonomy, Gibbs (1974) recognized as many as 23 groups.

All the indole alkaloids usually contain two nitrogen atoms; one indolic and the other generally 2 carbons removed from the β-position of the indole ring. These alkaloids have a restricted distribution. They are practically confined to families Apocynaceae, Asclepiadaceae, Rubiaceae (in these families their abundance makes them a family character!), Loganiaceae, Convolvulaceae and Fabaceae. Apart from angiosperms, they are also reported from fungi like ergot.

Catharanthus roseus G. Don (Apocynaceae)

Syn. *Vinca rosea* Linn.
Vinca pusilla Hook.f.
Lochnera rosea Reichb.

Baramasi, Vinca/Periwinkle

This plant is an erect, much-branched perennial with glossy oblong-elliptic leaves, fragrant white to pinkish purple flowers (blossoming in all seasons) in terminal or axillary cymose clusters and a pair of long slender follicles containing many oblong small seeds. A native of Malagasy, this plant is found wild or as cultivated throughout.

More than 110 indole alkaloids have been reported from this plant, but the more important are the 20 indole-indoline dimeric alkaloids such as vinblastine (vincaleukoblastine, VLB), vincristine (leurocristine, vincaleurocristine, VCR) that are mainly present in aerial parts. Other dimeric alkaloids present are vincarodine, vincoline, leurocolombine, vinamidine, vincathicine, vincubine and isositsirikine. Roots are the main source of alkaloids where the alkaloids amounts to 0.13%. Ajmalicine (raubasine), serpentine and reserpine (the principal alkaloids of *Rauwolfia*) are the main alkaloids, wherein their concentration exceeds that of *Rauwolfia*. Other important alkaloids are vincolidine, lochrovicine, catharanthine, vindoline, leurosine, lochnerine, tetrahydroalstonine, vindoline and vindolinine. Iridoids such as loganin, deoxyloganin, dehydrologanin, sweroside, secologanoside and roseoside form another group of compounds present in *C. roseus*. Also present are oleanolic acid and α-amyrin acetate. Flowers contain the alkaloids, coronaridine, 11-methoxy tabersonine, tetrahydroalstonine, ajmalicine, vindorosidine, vincristine, etc., alongwith α-amyrin, cycloartenol, lupeol, campesterol, β-sitosterol, petunidin, malvidin, hirsutidin, kaempferol and quercetin. Leaves contain flavonols such as quercetin and kaempferol, phenolic acids, vanillic, syringic and p-coumaric acids and ursolic acid. Seeds yield 31% of a fatty oil and loganic acid.

Vincristine, vinblastine, leurosidine and leurosine are oncolytic alkaloids effective against leukaemia cells. They are also used against Hodgkin's disease, Wilm's tumour, neuroblastoma, etc. The total alkaloids showed significant sedative and hypotensive action and cause relaxant and antispasmodic effect on smooth muscles and uterus. Alkaloids such as vincolidine, lochrovicine, etc., show diuretic activity. The alkaloid mixture is found to be hypoglycemic. The whole plant also exhibits antimicrobial activity.

Uncaria guianensis Gmeliu & *U. tomentosa* DC. (Rubiaceae)

Cat's claw

These are some of world's longest (upto 30m) woody vines with opposite simple elliptic to oblong leaves, curved strong stipular hooks in place of stipules, white flowers in clusters on a peduncle and elliptical capsules containing numerous small seeds. This is a native of Peru.

The stem/root contains 17 pentacyclic/tetracyclic oxindole alkaloids, anthraquinones, procyanidins and catechin tannins. Uncarine is the principal alkaloid.

The pentacyclic alkaloids show immune strengthening and anti-inflammatory activities, while the tetracyclic alkaloids affect CNS (and are toxic). The extract of the stem is used for arthritis, inflammations of the female urinary tract and speedy recovery after childbirth. They are also used to treat tumours, diabetes and gastric ulcers. Cat's claw is said to be a female contraceptive.

Strychnos nux-vomica Linn. (Loganiaceae)

Karaskara/Kuchla, Nux vomica

This is a tree of 40 meters or more, often armed with stiff spines. The leaves are ovate and shining. The small white flowers are borne in terminal compound cymes. The fruit is an orange-red berry with a hard shell and contains 3-5 discoid hard grey seeds. It is a native of India, Sri Lanka and Australia.

The seeds are the 'nux-vomica' of commerce. They contain 2.5-5% alkaloids, among which the prominent ones are strychnine (often amounts to 30-50% of the total alkaloids) and brucine (dimethyl strychine). Strychnine and brucine contain a quinoline skeleton in addition to indole nucleus—a feature which prompts some researchers to group them in quinoline alkaloids. Strychnine is concentrated towards the centre of the endosperm and brucine towards the periphery. The minor alkaloids accompanying are α- and β-colubrines, icajine, vomicine, novacine, N-oxystrychnine and pseudostrychnine. The seeds also contain chlorogenic acid, iridoids, and 3-4% of fixed oil. Loganin, one of the iridoids presumed to be an intermediate in the biogenesis of strychnine type alkaloids, occurs in high concentrations (upto 5%) in the fruit pulp. Other iridoids are 5-loganic acid, 7-O-acetyl loganic acid, 4-O-acetyl loganic acid, 6-O-acetyl loganic acid and 3-O-acetyl loganic acids.

This is a highly toxic plant, but in small doses, it is a cure for certain forms of paralysis and nervous disorders. Seeds are also used in chronic dysentery, rheumatism, insomnia and as an anthelmintic. Strychnine, being extremely toxic and a competent central stimulant, is an important tool in physiological and neuroanatomical research. It increases appetite and digestion and is also used as a tonic. It is occasionally used as a respiratory stimulant in certain cases of poisoning. Brucine is less toxic and used as an alcohol denaturant. This tree's bark is a febrifuge and tonic.

Strychnos ignati, a native of Philippines, is another commercial source of strychnine and has alkaloid content of 2.5-3% of which about 45-60% is strychnine and the rest brucine.

Physostigma venenosum Bal. (Fabaceae)

Calabar Bean

This native of W. Africa is a perennial woody climber with pinnate leaves of 3 ovate leaflets and numerous flowers in pendulous racemes. The legume is about 15 cm long and contains 2-3 flattened hard reniform seeds.

Seeds yield about 0.15-0.4% alkaloids of which physostigmine (eserine) is the major alkaloid. Eseramine, isophysostigmine and physovenine are the minor alkaloids accompanying physostigmine.

Physostigmine—a very easily oxidizable alkaloid that needs to be protected from air and light—is a reversible inhibitor of cholinesterase and used extensively in opthalmology for contracting the pupil of the eye. The other alkaloids also possess this activity, but only physovenine possesses the same intensity of action.

Mucuna pruriens DC. (Fabaceae)

Syn. *M. prurita* Hook.

Atmagupta/Kavach, Mucuna

A twining annual with pubescent branches, this species has pinnately trifoliate leaves with ovate leaflets and large purple flowers on drooping racemes. The fruit is an S-shaped legume densely clothed with brown, stinging hairs. It is a native of India and grows in many parts of S.E. Asia.

Seeds and roots of this plant contain a variety of alkaloids such as physostigmine, mucunine, mucunadine and nicotine along with 3-4% of a red viscous oil and DOPA (Dihydroxy phenylalanine). The hairs on the fruit contain 5-hydroxy tryptamine, which is said to be responsible for the irritation caused by them.

The roots are used against cholera, as a diuretic and for dropsy. Seeds are powerful aphrodisiacs and are known to promote sexual vigour and semen. They are also used in cholera, delirium, impotence, leucorrhoea, and spermatorrhoea and as a diuretic. It is also found to exert an antidepressant effect in cases of depressive neurosis.

Claviceps purpurea Tul. (*Clavicipitaceae*)

Ergot

The sclerotia produced by the parasitic fungus on the ovary of many grasses, especially rye, are the ergot of commerce. The infection takes place in spring and the mycelium penetrates the ovary. These mycelia feed on the ovarian tissue and finally replace it with a compact "pseudoparenchymatous" tissue, the sclerotium or resting stage. The sclerotium increases in size in summer and projects out of the inflorescence as finger-shaped protruberances.

The sclerotia, which form the drug, are dark violet in colour, slightly curved fusiform structures 1-5cm long. Ergot yields 0.15-0.20% of total alkaloids. Six pairs of alkaloids are recognized which are grouped into two classes based on their solubility in water. The water-soluble ergometrine (ergonovine) group includes ergometrine and ergometrinine. In ergometrine, lysergic acid is linked with L-2-amino propan-l-ol by an amide bond. These are also called clavine alkaloids. The water-insoluble groups are ergotamine group (ergotamine, ergotaminine, ergosine and ergosinine) and ergotoxine group (ergocristine, ergocristinine, ergocryptine, ergocornine and ergocorninine). These alkaloids are known as peptide alkaloids because within these groups, lysergic acid or its isomer is linked with a peptide, e.g. in ergotamine, the carboxyl group of lysergic acid is linked to a peptide which, on hydrolysis, yields pyruvic acid, ammonia, L-phenylalanine, and D-proline (in intact alkaloid proline occurs in L-form). On hydrolysis, the ergotoxine group yields lysergic acid, dimethyl pyruvic acid, ammonia, D-proline and one of the following amino acids, i.e. L-leucine (in ergocryptine), L-phenylalanine (ergocristine) or L-valine (ergocornine). In all these alkaloids, the amino acids occur in the form of a cyclic peptide.

In addition to these alkaloids, ergot contains several pigments, fixed oil (upto 35%), sterols like ergosterol (fungisterol), histamine, tyramine and acetylcholine.

Ergot is used for its oxytocic ("quick delivery") effect, a property due to ergometrine. Ergometrine is also used for preventing and treating post-partum haemorrhage. Ergotamine and ergotoxine are specific analgesics for migraine. Ergolines are potential prolactin and mammary tumour inhibitors. Lysergic acid diethylamide (synthetic)-LSD 25 is a hallucinogen and a potent psychotomimetic.

The other fungi which are found to synthesize ergot alkaloids (especially clavine alkaloids) are *Aspergillum fumigatus, Penicillium chermesinum. P. roquefortii* and *Rhizopus arrhizus*. Some of the Convolvulaceae members also are found to possess the same alkaloids. The important ones are: (1) *Rivea corymbosa* Hall. (chanoclavine, elymoclavine, lysergol and ergine), (2) *Argyreia nervosa* Boj, and (3) *Ipomoea purpurea* Roth (both containing ergine, ergometrine, ergometrinine, isoergine and penniclavine). The maximum yield reported is 0.05%.

Convolvulus pluricaulis Chois. (Convolvulaceae)

Syn. *Convolvulus microphyllus* Sieb.
C. prostratus Forsk.

Shankhapushpi

This is a procumbent perennial with a thick woody root stock, slender, wiry and thinly hairy spreading stem, small subsessile, spatulate villous leaves, white funnel-shaped flowers, either singly or in groups of 2-3 and a small globose capsule. This occurs as a weed in western India.

The plant is found to contain clavine alkaloids such as convolvine, convolamine, phyllabine, convolidine, contoline, convoline, subhirsine and convosine. Other compounds reported are cerylalcohol, β-sitosterol, 6-methoxy coumarin, 4'-methoxy kaempferol, 7-methoxy quercetin, phenolic acids like p-hydroxy benzoic, vanillic, syringic and melilotic acids, quinones, β-sitosterol and microphyllic acid (Nair et al. 1988a).

The whole plant is used as hypotensive, sedative, in anxiety neurosis and for improving mental function. It is found to reduce different types of stress including psychological, chemical or traumatic. It also acts as an antithyroid.

Evolvulus alsinoides Linn. (Convolvulaceae)

Vishnukranta/Shankhapushpi

A prostrate perennial herb with a small woody branched rootstock, numerous wiry spreading branches clothed by long-spreading hairs, small elliptic, oblong, almost sessile leaves, densely clothed by silky hairs, small blue flowers, solitary or in pairs from the axils of leaves, and globose four-valved capsules containing four seeds. This weed is common in tropical and subtropical countries.

The whole plant yields an alkaloid evolvine, pentatriacontane, triacontane β-sitosterol, a glycoflavone, 4'-methoxy vitexin and phenolic acids such as p-hydroxybenzoic,vanillic, protocatechuic and gentisic acids and quinones (Nair et al. 1988b).

The entire plant is a specific remedy for all kinds of fevers. It is also a powerful brain stimulant, aphrodisiac, and has curative powers for insanity, epilepsy, nervous debility and dysentery. The flowers are good remedy for uterine bleeding and the roots for gastric and duodenal ulcers. The entire plant is used to treat leucoderma and its leaf juice in conjunctivitis.

Ipomoea nil Roth. (Convolvulaceae)

Syn. *I. hederacea* Jacq

Krishnabijah

I.nil is a twiner with a hairy stem, ovate-cordate 3-lobed hairy leaves, large showy funnel-shaped blue flowers, either solitary or in few flowered axillary cymes and a subglobose capsule containing black glabrous seeds. A native of America, it now grows in all regions.

The leaves of this plant contain flavonoids, acacetin and 7-methoxy luteolin and phenolic acids *p*-hydroxy benzoic, gentisic, vanillic, syringic and ferulic acids. The seeds are found to contain indole alkaloids (lysergic acid derivatives) chanoclavine, penniclavine, isopenniclavine and elymoclavine, and lysergol (Nair et al. 1988b)

Seeds, the official drug, act as a good purgative, anthelmintic, blood purifier and are used in rheumatism, paralytic affections and as a paste for skin diseases. The plant extract is hypoglycemic.

Ipomoea mauritiana Jacq. (Convolvulaceae)

Syn. *I. digitata* Linn.

Vidarikanda/Vidari

Vidari is a large laticiferous glabrous twiner with ovoid or elongated tuberous roots, long thick stems, long-petioled, broad palmately 5-7 lobed leaves, large purple campanulate flowers in corymbosely paniculate cymes and an ovoid capsule containing four ovoid woolly seeds. Found throughout India, this plant is more common in sandy shores/banks.

The tuber of vidari is found to contain β-sitosterol and a glycoside paniculatin. The leaves contain 4'-methoxy quercetin, 3',4'-dimethoxy quercetin, gentisic acid, vanillic acid and syringic acid (Nair et al. 1988a)

Vidarikanda is a sweet-tasting restorative tonic, aphrodisiac, galactogogue, diuretic and demulcent. It promotes strength and complexion. Ether-soluble fraction is hypotensive and a muscle relaxant. The leaves are applied externally for rheumatism.

Ipomoea marginata Verdc. (Convolvulaceae)

Syn. *I. sepiaria* Koenig.

Lakshmana

This medicinal plant is a perennial twiner with ovate-cordate acute leaves having reddish patches, light pink (having a dark eye), funnel-shaped flowers in pedunculate subumbellate cymes and ovoid glabrous capsules containing 2/4 grey seeds with silky pubescence. It is very common near water bodies throughout India.

The leaves contain 4'-methoxy quercetin and 3',4'-dimethoxy quercetin, proanthocyanidins and phenolic acids *p*-hydroxy benzoic, gentisic, protocatechuic, vanillic and ferulic acids (Nair et al. 1988a).

In Ayurveda, Lakshmana is the only plant which can cure the sterility in women. This is also believed to have the power of bestowing a male child. Its root is the official part for promoting bodily strength and also acts as a tonic.

Erythrina variegata Linn. (Fabaceae)

Syn. *Erythrina indica* Lamk.

Paribhadrah, Coral tree

E. variegata is a large deciduous tree armed with small conical prickles, and has stellate pubescent branches, trifoliolate leaves bearing rhomboid-ovate inequilateral leaflets, bright red flowers in clustered racemes at the tips of branches and a cylindrical turgid pod compressed between the seeds. Seeds are four to eight, and subreniform. Paribhadrah is common in coastal areas and in forests of the Western Ghats.

The stem bark of this tree contains indole alkaloids such as erysotrine (major), erysodine, erysovine, erysonine, hypaphorine and N-N-dimethyl tryptophan. Also present are stachydrine, wax, alkyl ferulates, phenolates and flavones, osajin, alpinum isoflavone, oxyresveratrol and erythrinins A, B, & C. The leaves also contain erysotrine (major), erysodine, erythraline, hypaphorine, erythrinine, erysodine and de-N-Me-orientaline along with vitexin, isovitexin, proanthocyanins and melilotic acid. The red flowers yield erysotrine (major), erythrartine (ll-OH erysotrine), hypaphorine and choline. The seeds contain same alkaloids (erysodine, hypaphorine, erysopine and erysotrine, erythraline and erysovine), a fatty oil and lectins.

Paribhadrah (bark and leaves) is anthelmintic, carminative, diuretic, galactogogue, expectorant and febrifuge. It is also used in anorexia, obesity, dysmenorrhoea, rheumatism and skin diseases.

Rauwolfia serpentina Benth. (Apocynaceae)

Sarpagandha, Rauwolfia

The dried roots and underground portions of the stem of *R. serpentina,* yield the drug rauwolfia. It is a laticiferous shrub, 1-1.5 m, in height with simple ovate leaves in clusters of 3-5 at a node, pale rose white flowers in terminal or axillary cymes and a purplish black drupe. Rauwolfia is a native of India.

The roots yield about 0.7-1.5% alkaloids consisting of at least 50 compounds. These alkaloids differ in basicity, possessing weak (simple indoles) bases, strong bases (anhydronium bases) and also intermediate basicity (indoline alks). Of the principle alkaloids, reserpine, rescinnamine and deserpidine are yohimbane type possessing a pentacyclic skeleton where the fifth ring (E) is carbocyclic, whereas ajamalicine, tetrahydroreserpine, raubasine and reserpinine exhibit a heteroyohimbane skeleton—the fifth ring is heterocylic. Serpentine, serpentinine and alstonine are strongly basic anhydronium alkaloids. The other constituents of the roots are a phytosterol, fatty acids and sugars.

Here, only reserpine–rescinnamine group of alkaloids are therapeutically important. They exhibit antihypertensive and tranquilizing action. Reserpine is used effectively against hypertension. The sarpagandha root is used in cases of high blood pressure, insanity and schizophrenia. It is also used as anthelmintic, a cardiac depressant, for insomnia, epilepsy and snake bites. Decoction of the root causes uterine contractions and promotes expulsion of foetus.

The related species yielding the above-mentioned alkaloids are: (1) *R. vomitoria* Afzel (of tropical Africa); and (2) *R. tetraphylla,* L. (of India, containing more of deserpidine, which possesses the same activity of reserpine but without the side effects).

Passiflora incarnata Linn. (Passifloraceae)

Krishna kamal, Passionflower

P. incarnata is a perennial climber with three-lobed finely serrated leaves bearing axillary tendrils, buff-coloured (tinged with purple corona), sweet-scented flowers and orange coloured, ovoid, many-seeded berries containing edible sweet pulp. This vine is a native of America.

The plant contains alkaloids such as harmine, harman, harmaline, harmol and harmalol, flavonoids like apigenin, luteolin, quercetin and rutin, sterols and gums.

The aerial parts are used for their tranquilizing action in anxiety neurosis, insomnia and for improving concentration, an activity attributed to flavonoids. The drug also possesses antispasmodic and analgesic tendencies.

Alstonia scholaris R. Br. (Apocynaceae)

Saptaparna, Dita bark tree

A buttressed, large, evergreen laticiferous tree with 4-7 ovate leaves in a whorl, with greenish white fragrant flowers in umbellate cymes and a pair of long slender follicles containing brown comose seeds, *A. scholaris* is a native of S. Asia and Africa.

The stem bark is dark grey (when young) to brownish buff and contains alkaloids (upto 2.7%) with echitamine as the major component and echitamidine, tubotaiwine, picrinine, akuammidine, ditamine and echitenine as minor components. Other constituents of stem bark are α-amyrin acetate and lupeol acetate. The alkaloids present in the root bark are demethylechitamine, akuammicine, tubotaiwine and pseudoakuammigine. The triterpenoids presents are α-amyrin, lupeol, stigmasterol, β-sitosterol and campesterol. Leaves yield alschomine, isoalschomine, tubotaiwine, lagunamine, 19-isoepischolaricine, scholaricine, picrinine, pseudoakuammigine, picranilal, β-sitosterol, netulin and ursolic acid. The latex of the plant is found to contain caoutchouc (upto 8%) and resins besides certain proteins. Flowers yield a volatile oil and four alkaloids: picrinine (major), strictamine, tetrahydroalstonine and an unidentified alkaloid as well as lupeol, β-amyrin and ursolic acid. The constituents of the volatile oil are α-pinene, Δ^3-carene, limonene, terpeolene, linaool (major, 30%) linalyl acetate and citral. Fruits contain only akuammidine (Yamauchi et al. 1990).

Almost all parts of the plant are used in medicine. The bark is an astringent, anthelmintic and antiperiodic. It is used against chronic diarrhoea, dysentery and bowel complaints. Echitamine and also the total alkaloids cause a sharp fall in blood pressure and are hypotensives. These compounds are found to be active against human sarcoma. Leaves are used against beriberi, congestion of liver, dropsy and ulcers. The latex is useful as an application for ulcers, sores, and tumours and also in rheumatic pain.

Peganum harmala Linn. (Zygophyllaceae)

Harmal, Peganum

The seeds of *P. harmala,* a native of W. Asia and India, have been known for their psychotropic effect. This plant is a dichotomously branched bush of 1 m in height with a dense foliage consisting of leaves having pinnately cut linear, narrow, acute spreading lobes, small solitary axillary white flowers and globose capsule enclosing numerous angular seeds.

The seeds contain β-carbolines like harmine (the main constituent), harmalol and harman.

The active hallucinogen is harmine, which exhibits greater activity than mescaline. The seeds find use as a spice and are a valuable aphrodisiac.

Desmodium gangeticum DC. (Fabaceae)

Syn. *Hedysarum gangeticum* Linn.

Prasniparni, Desmodium

A diffuse undershrub with irregularly angled stems, unifoliolate, stipulate leaves having ovate-acute membranous leaflets clothed with dense soft whitish appressed hairs, small pink flowers in ascending terminal/axillary racemes and subfalcate deeply indented (on the lower edge) moniliform lomentum, sparsely clothed with minute hooked hairs. Desmodium is a common weed in all tropical countries.

Aerial parts contain tryptamine derivatives and alkaloids such as N-Me-tetrahydroharman and 6-OMe-2-Me-β-carbolinium cation. Roots yield N, N-dimethyl tryptamine and its N-oxide, N-methyl tyramine, hypaphorine, hordenine and candicine alongwith pterocarpans like gangetin and desmodin. Seeds also are found to contain β-carbolines and indole-3-alkylamine.

The root which forms one of the components of *dasamoola* (ten roots) is a good cardiotonic and is used in cardiac disorders. It is also a diuretic, laxative, nervine tonic and useful in treating insanity, ulcers, dysentery and fever. Total alkaloids from aerial parts show hypotensive activity.

Murraya exotica Blance (Rutaceae)

Murraya

This is a native of Australia. This small attractive tree bears shining pinnate leaves of 3-9 ovate leaflets. The flowers are small, white, and borne on panicles. The fruit is an ovoid red berry.

The bark yields a series of alkaloids based on carbazole skeleton, prominent among them being murrayanine and girinimbine. The minor alkaloids present are mahanimbine, carrayangin, mahanine and mukoeic acid. The bark contains also a coumarin, mexoticin. The leaves yield a volatile oil and flavonoids such as 7,4'-diOMe vitexin. 4'-OMe kaempferol, quercetin, and phenolic acids such as vanillic, syringic and *p*-coumaric acids (Umadevi et al. 1990).

The bark is used as a stimulative tonic and possesses antidiarroheal properties. The leaves are extensively used to flavour food and for the treatment of nausea, biliousness, dyspepsia and diarrhoea.

M. koenigii Spreng, a related plant, is extensively used as a leaf spice in India. It is found to promote appetite and digestion and is used in indigestion, acidity, and diarrhoea and also acts as an anthelmintic. The leaves are found to contain 7,4'-diOMe vitexin, 4'-OMe kaempferol, vanillic acid, syringic acid and *p*-coumaric acid (Umadevi et al. 1990).

Gelsemium sempervirens Ait. (Loganiaceae)

American Yellow Jasmine

This evergreen glabrous twining shrub bears lanceolate leaves, fragrant yellow flowers in cymes and flat capsular fruits. It is a native of America.

The dried rhizomes and roots contain about 0.15-0.25% alkaloids. Gelsemine, gelsemicine, and sempervirine occur in larger quantities and gelsedine and gelsevirine in traces. All these alkaloids except sempervirine belong to oxindole group. The roots yield scopolin (coumarin).

The drug being a drastic CNS depressant, is used with extreme care.

Tabernaemontana divaricata Roem. &Schult. (Apocynaceae)

Syn. *T.coronaria* Willd.
Ervatamia divaricata Burkill.

Nandyarvatah/Tagar

A glabrous, much-branched laticiferous shrub with shining oblanceolate-acuminate leaves and large white fragrant flowers (corolla twisted, corona absent). The fruits of this plant have not been seen. A native of north India, it is cultivated in the gardens of tropical countries.

Leaves are found to contain flavonols kaempferol and quercetin and phenolic acids, salicylic, sinapic, syringic, vanillic and *p*-hydroxy benzoic acids (Daniel and Sabnis, 1978).

The flowers and buds are useful in treating eye and skin diseases. The roots are anthelmintic, anodyne, tonic and used for rheumatism. The latex is used to heal wounds.

2.5 Isoquinoline Alkaloids

Numbering more than 400, these alkaloids, possessing an isoquinoline nucleus, form a large class by itself. The three common ring structures, i.e. isoquinolines, benzyl isoquinolines and phenanthrenes and their various modifications and derivatives make this a very complex assemblage necessitating further classification. Out of the many classification systems, the one proposed by Gibbs (1974) appears to be comparatively simple and convenient, wherein 12 groups of isoquinolines are recognized.

1. Simple Isoquinolines

There are about 20 alkaloids possessing a tetrahedroisoquinoline skeleton only (e.g. anhalamine from *Lophophora williamsii*–Cactaceae).

2. Benzyl Isoquinolines

These alkaloids have a benzyl group attached to the position 1 of isoquinoline nucleus. There are about 60 representatives (e.g. Papaverine from *Papaver somniferm* Linn.).

3. *bis*-Benzyl Isoquinolines

These are dimers of benzyl isoquinolines wherein two benzyl isoquinoline nucleii are joined by one, two or three ether linkages. They are believed to have originated through enzymative dehydrogenations of bisbenzyl isoquinoline units. Known structures are about 80. (e.g. Tiliacorine from *Tiliacora racemosa*–Menispermaceae).

4. Aporphines

Aporhines arise naturally by oxidation of phenols of the benzyl tetrahedro-isoquinolines. Known members are about 100, mostly restricted to Magnoliales and Ranunculales (e.g. Glaucine in *Thalictrum minus* Linn.–Ranunculaceae).

5. Cularines

Cularines are also derivatives of benzyl isoquinolines in which a phenolic oxygen is incorporated into a seven-membered ring. A very small group of 4 compounds is reported from *Dicentra* and *Corydalis*, e.g. Cularine, *Dicentra cucullaria* Bernh. (Papaveraceae).

6. Morphines

Morphines possess a phenanthrene nucleus. Numbering about 20, they are reported mostly from Papaveraceae (e.g. Codeine from *Argemone mexicana* Linn.).

7. Protoberberines

There are about 50 members in this group, derived by condensation of benzyl isoquinolines and formaldehyde molecules. They are frequent in Papaveraceae and reported from many families of Magnoliales and Ranunculales (Berberine from *Berberis*.).

8. Protopine Group

An isoquinoline nucleus is absent in this group of alkaloids. They are included in isoquinolines because they are derived from isoquinolines and are reconvertible. Protopines are characterized by the presence of a 10-membered heteroring containing one carbonyl group. Only 14 compounds, mostly from Papaveraceae, have been isolated so far. (e.g. Protopine).

9. Phthalide Isoquinolines

They are derivatives of benzyl isoquinolines wherein a furanolactone ring is attached to the benzyl group. Only 11 alkaloids are reported so far, all from Papaveraceae (e.g. narcotine from *Papaver).*

10. Emetine Group

This group is characterized by the presence of both isoquinoline and quinolizidine nucleii. All the 17 compounds known are isolated from Rubiaceae (e.g. Emetine from *Cephelia ipecacunanha* Rich).

11. α-Naphthaphenanthridines

These alkaloids possess tetracyclic ring systems probably derived from tetrahydroberberine systems. About 18 alkaloids have been isolated mostly from Papaveraceae (e.g. Chelidonine from *Chelidonium majus* Linn.).

12. Miscellaneous

The rest of the isoquinolines—about 6, which cannot be grouped in the above classes—are placed in this group, e.g. Thalicarpine, a benzyl isoquinoline-aporphine alkaloid from *Thalictrum dasycarpum* Fisch. (root).

Cephaelis ipecacuanha A. Rich. (Rubiaceae)

Syn. *Psychotria ipecacuanha* Stokes

Ipecac

Ipecac is a small straggling woody herb with slender rhizomes bearing long annulated roots (brick red/brown with a thick bark showing transverse furrows), striated stems, oblong/elliptic leaves, white flowers arranged in heads and clusters of dark purple berries enclosing two plano-convex seeds. A native of Brazil, this plant is now cultivated everywhere.

The roots, which form the drug, contain about 2% alkaloids, of which emetine amounts to 60%. The other two major alkaloids present are cephaeline and psychotrine (together about 35%). Also present are tetrahydroisoquinoline-monoterpene glycosides-6-O-methyl ipecoside, ipecosidic acid, demethyl alangiside, neoipecoside, 7-O-methyl ipecoside, 3,4-dehydroneoipecoside and *cis* and *trans* cephalosides. The minor alkaloids reported are O-methyl psychotrine, emetamine, ipecamine, hydroipecamine, protoemetine and kryptonine. In addition to alkaloids, the drug contains glycosides ipecoside, alangicide, sweroside and ipecacuanhin, erythrocephaelin (a red pigment), dehydrologanin, choline, ascorbic acid, D-mannitol, tannins, organic acids like malic and citric acids, saponin, resin, calcium oxalate and starch. The stem, leaves, flowers and fruits also contain same alkaloids but with varying amounts (Itoh et al. 1991; Nagakura et al. 1993).

Ipecac is used as an expectorant, emetic and for amoebic dysentery. Emetine has more expectorant activity and cephaeline more emetic.

Berberis aristata DC (Berberidaceae)

Daruharidra, Barberry

B. aristata is a spiny shrub with obovate/elliptic toothed leaves, sharp spines from the nodes, yellow flowers in corymbose racemes and oblong-ovoid red berries. It is a native of Nepal, widely cultivated or grown as an escape in many parts of western India.

The root bark, as well as stem bark, which forms the drug, yields (upto 4%) isoquinoline alkaloids, such as berberine, berbamine, aromoline, karachine, palmatine, oxycanthine, oxyberberine and taxilamine.

The total alkaloid extract is a bitter tonic, used as a cholagogue, laxative, diaphoretic and applied in eye infections, ulcers and hemorrhoids. It is also an emmanagogue and is useful in the treatment of jaundice. The fresh berries are laxative, antiscorbutic and useful in piles and eye diseases, especially conjunctivitis.

B. asiatica Roxb., common in the Himalayas, also yields similar alkaloids (berberine amounts of 2% in root bark) and is used in place of barberry. Related species such as *B. chitria, B. coriaria, B. floribunda,* etc., also are used as substitutes.

Alangium salvifolium Wang (Alangiaceae)

Syn. *A. lamarkii* Thw.
A. decapetalum Lamk.

Ankol, Alangium

This medium-sized tree of Indo-Malayan origin possesses ovate/oblong acuminate leaves, white scented flowers (with 10 linear petals) and ovoid tomentose fruits with persistent calyx, containing large seeds enclosed in red mucilaginous pulp. The plant grows in dry regions all over India.

All parts of the plant are rich in isoquinoline alkloids. The alkaloids of root bark are alangine A & B, desmethyl psychotrine, marckine, markidine, lamarkinine, psychotrine, tubulosine, isotubulosine, cephaeline and emetine. Protoemetinol, alangisterol, stigmasterol, β-sitosterol and a fatty oil with myricyl alcohol and fatty acids such as oleic and palmitic acids are the other components of the root bark. The roots contain fewer alkaloids: cephaeline, tubulosine, isotubulosine, psychotrine and alangicide. The stem bark also is rich in alkaloids: cephaeline, psychotrine, tubulosine, desmethyl psychotrine and demethylacephaeline as well as β-sitosterol, stigmasterol and oil. The leaves contain alangimarkine, ankorine, deoxytubulosine, alangicide and 3'-*epi*-tubulosine as their alkaloidal constituents alongwith stigmaste-5, 22, 25-trien-3-β-ol, β-sitosterol, friedelin, N-benzoyl-L-phenyl-alaninol, triterpene A, isoalangidiol, alangidol, myristic acid and choline. The fruits yield alkaloids such as cephaeline, N-methyl cephaeline, deoxytubulosine and alangicide. The seeds of this tree contain emetine, cephacetrine, psychotrine, N-methyl cephaeline, alangicide and alangimarkine. Terpenoid components of seeds are belulinic acid, betulinadehyde, betulin, lupeol, hydroxylactone-A, belulinic acid and β-sitosterol.Other constituents reported from the plant (root, leaves and fruit) are alangicide (a monoterpenoid lactam), loganic acid, venoterpine, *dl*-salsoline and isocephaeline.

The root and root bark of Ankol are astringent, anthelmintic, purgative, emetic and useful in skin diseases. It is a reputed single drug in Ayurveda for treatment of rabies. It is a substitute of ipecac and exhibits anti-tubercular activity and is antipyretic, diaphoretic and administered as a cure for syphilis. Roots are hypotensive in action. Stem bark possesses alkaloid mixture, AL-60, and exerts marked hypotension in cats. The alkaloid extract of leaves is found to be hypoglycaemic, antispasmodic and known to exhibit anticholinesterase activity. Fruits are anti-phlegmatic, laxative, tonic and refrigerant and useful in emaciation and haemorrhages. Even the seeds are cooling, tonic and useful in haemorrhages. The alkaloid extract of the seeds causes hypotensive (higher doses) and hypertensive (lower dose) effects on rats.

Hydrastis canadensis Linn. (Ranunculaceae)

Hydrastis, Golden Seal

Indigenous to Canada and the United States, this erect perennial has a yellow rootstock. The number of leaves is normally three, of which one is long and radical and other two round, 5-7 lobed and cauline.

The drug, formed by the rhizome and the numerous slender roots, contains 2.5 to 4% ether soluble alkaloids of which the principal one is hydrastine. Berberine, canadine and berberastine (5-hydroxy berberine) are present in minor quantities.

Hydrastis finds extensive use in checking uterine haemorrhage. Hydrastine salts also exhibit this property. The drug is used as an astringent in inflammation of mucous membranes.

Mahonia aquifolium Nuttal (Berberidaceae)

Berberis

An evergreen shrub, 1 m in height, Berberis has pinnately divided leaves with 5-9 ovate/lanceolate glossy leaflets. The yellow flowers are arranged in fascicled racemes. The fruit is a blue glaucous berry. Berberis is indigenous to Rocky Mountains in the USA.

The dried rhizomes and roots yield a number of alkaloids. The amount of berberine is always very high compared to other alkaloids such as oxycanthine, berbamine, isotetrandrine and magniflorine. Berberis is employed as a bitter tonic, alterative, stomachic and a curative for piles. Berberine is an effective antipyretic and antiperiodic. It is also used as a medicine for erectile dysfunction.

Berberis forms the main source from which berberine is commercially extracted. Other species like *Berberis vulgaris* Linn,. *B. asiatica* Roxb. And *B. aristata* Linn. are also exploited as sources of berberine, of which the amount of berberine in *B. asiatica* often exceeds 2%.

Argemone mexicana Linn. (Papaveraceae)

Swernakshiri/Bharband, Mexican Poppy

This herbaceous annual bears sessile, pinnately lobed leaves with toothed margins. The beautiful yellow flowers occur singly on erect peduncles. The capsule is prickly, containing numerous seeds resembling mustard. A native of Mexico and S. America, *Argemone* occurs as a weed everywhere now.

The leaves and stem contain a number of alkaloids like berberine, protopine, retundine, munitagine, argemonine, nor-argemonine, codeine, morphine, toddaline, sanguinarine, etc. Most of these alkaloids are present in the seeds also. Apart from alkaloids, seeds yield 15-30% fixed oil.

The bitter yellow juice of the leaves and stems is used in chronic opthalmia. Its seeds are narcotic, and a substitute for ipecacaunha. The seed oil is used in relieving the bowels—wherein sanguinarine occurs as a carcinogenic contaminant.

Sanguinaria canadensis Linn. (Papaveraceae)

Bloodroot

A native of N. America, bloodroot contains orange-red latex in its horizontal thick branching rhizome. The one-flowered scapes, bearing white flowers, are often overtopped by the solitary palmately lobed basal leaf. The fruit of this plant is an ellipsoidal capsule enclosing numerous seeds.

The alkaloid content of rhizomes reaches 5% or more. The major alkaloids are sanguinarine (about 1%), chelerythrine, protopine and allocryptopine—all of protopine series. Sanguinarine and chelerythrine, although colourless, form red and yellow salts, respectively. The other constituents of the drug are a red resin and a good amount of starch.

Bloodroot is a stimulating expectorant and emetic, while sanguinarine causes doubling of chromosome in dividing calls, a property similar to colchicine.

Curare

The term **curare** denotes a variety of South American arrow poisons extracted by the natives from various plant sources. The plants often used are *Chondrodendron* spp, i.e. *C. tomentosum* R & P., *C. microphylla* and *Anomospermum grandiflora* (Menispermaceae) and spp. of *Strychnos,* i.e. *S. castelnaea, S. toxifera* and *S. crevauxii* (Loganiaceae). New sources are continuously discovered in course of time. Depending upon the containers in which they are available, the curares are named tubecurare (in bamboo tubes), calabash curare (in gourds) and pot curare (in earthernware pots).

The crude dried extract obtained after boiling the bark with water is the curare of commerce. It contains a variety of alkaloids and quaternary compounds that differ, depending on the source plants. Menispermaceous curare contains (+) tubocurarine chloride—a quaternary compound containing a bisbenzyl isoquinoline structure—as the principal alkaloid and four non-quaternary bases—isochondrodendrine, isochondrodendrine dimethyl other, berberine and chondrocurine. Loganiaceous curare contains mostly C_{40} compounds of dimeric strychnine type (indolic) designated as toxiferines and C-curarines.

Crude extract exhibits a paralyzing effect on voluntary muscle and is toxic to blood vessels. Tubocurarine chloride is used as a skeletal muscle relaxant and to control convulsions caused by strychnine poisoning and tetanus.

Tinospora cordifolia Miers (Menispermaceae)

Amruta/Guduchi, Tinospora

T. cordifolia is a woody twiner often producing filiform green aerial roots. The stem has a warty surface (due to lenticels) and the leaves are thick, heart shaped and acuminate. The flowers are small, unisexual, borne on racemes in the axils of leaves, while the fruit is a globose drupe. This plant is a native of India, common in tropical and subtropical parts.

The stem, which forms the drug, contains clerodane furano diterpenes, viz., columbin and tinosporaside; a lignan 3,4-*bis* (4-hydroxy-3-methoxy benzyl) tetrahydrofuran; alkaloids such as jatrorrhizine, palmatine, berberine and timbeterine; a sesquiterpene glycoside, tinocordifolioside; phenylprepene disaccharides cordifolioside A & B; choline, tinosporic acid and tinosporon (Pachey and Schneidir, 1981; Wazir et al. 1995; Maurya et al. 1996, 1997).

Amruta is one of the rejuvenating drugs useful in general debility, rheumatism, diabetes, and improvement of intellect and a cure for fever. It also exhibits immunosuppressive (Pendse et al. 1997), antiinflammatory, analgesic and antihepatotoxic properties (Singh et al. 1984).

Cissampelos pareira Linn. (Menispermaceae)

Abuta, False Pareira

Abuta is a slender dioecious twiner with a woody rootstock, peltate or orbicular reniform leaves having a truncate/cordate base, small greenish yellow unisexual flowers (male flowers in axillary fascicled cymes, female flowers in long pendulous racemes) and small, ovoid, scarlet red drupes containing horse shoe-shaped seeds. A common climber along hedges and waste places, abuta is a native of S.E. Asia.

The roots contain a number of bisbenzyl isoquinoline alkaloids, a volatile oil, a fixed oil and quercitol. The alkaloids located are hayatine (± bebeerine/curine/chondocurine), hayatinine (4"-methyl bebeerine), hayatidine, *l*-bebeerine (pelosine /cissampeline), isochondrodendrine, menismine, pareirine, cyclanoline, cyclearine, dehydrodicentrine, dicentrine and insularine. Leaves and stems contain alkaloids, laudanosine, nuciferine, bulbocarpine, corytuberine, magnoflorine, grandirubine, isomerubrine and tropoloisoquinolne alkaloids, pareirubines A&B and azafuranthene alkaloids norimeluteine and norrufescine. The volatile oil (upto 0.2%) contains thymol as the main ingredient. The leaves are found to contain *l*-bebeerine, cycleanine, hayatidine, hayatinine, hayatin and cissampeloflavone (a chalcone-flavone dimer).

All parts of the plant exhibit medicinal properties. The roots possess diuretic, antilithic, analgesic, antipyretic and emmanagogue properties and prevent threatened miscarriage. They are used in dysentery, piles, dropsy and urinogenital troubles such as uterine prolapsis, cystitis, haemorrhage and menorrhagia. Hayatin methiodide and methochloride are neuromuscular blocking agents. Cissampeline is active against human carcinoma cancer cells. Pareirubines A & B are antileukemic.

Stephania glabra Miers. (Menispermaceae)

Stephania

This native of Asian tropics is a woody climber possessing large tuberous roots and broadly ovate leaves. Flowers are unisexual in axillary umbels. The fruit is a reddish drupe.

Roots yield a wide variety of isoquinoline alkaloids like gindarine (palmatine), gindaricine, gindarinine (all protoberberines), pronuciferine (benzyl isoquinoline), dicentrine (aporphine type) homostephanoline and hasubanonine (both morphine type). The roots are extensively used in pulmonary tuberculosis, asthma, dysentery and fever.

Jateorhiza palmata Linn. (Menispermaceae)

Calumba Root

The roots of this dioecious climber are large fusiform, growing from a rhizomatous base and yield about 2-3% alkaloids consisting of protoberberines like palmatine, jateorhizine, calumbamine and a dimer, *bis*-jateorhizine. Also present in the drug are diterpenoid bitter principles, calumbin, chasmanthin and palmarin.

Calumba root is used as a bitter tonic, analgesic and for diarrhoea and dysentery.

Tiliacora racemosa Cole. (Menispermaceae)

Tiliacora

This twining shrub bears glabrous ovate leaves, fragrant cream coloured unisexual flowers arranged in axillary racemes, and a drupaceous fruit. It is a native of India.

The leaves of this plant contain the alkaloids tiliacorine (a bisbenzyl isoquinoline with a diphenyl grouping), tiliacorinine (the diasteroisomer of tiliacroine), *nor*-tiliacorinine and tiliacoridine. Root bark contains tiliarine, a stereoisomer of nor-tiliacorine.

The leaves are used as an antidote for snake bites and for filariasis. Roots are used against skin infections. Total alkaloidal fraction showed marked hypotensive activity.

The other important menispermaceous plants are: (1) *Cocculus pendulus* Diels., containing six biscoclaurine alkaloids of which cocculinin shows anticancer properties, and (2) *C. laurifolius* DC, yielding isococculidine, an alkaloid possessing powerful neuro-muscular blocking action.

Papaver somniferum Linn. (Papaveraceae)

Afim, Opium Poppy

Probably a native of Asia Minor, opium poppy is a glabrous annual attaining about 1 m height. Leaves are variable. The lower leaves are narrowed to a short petiole whereas the upper ones are stem clasping, cordate and lobed. The large, pinkish-white flowers are borne solitary, on long peduncles. The fruit is a capsule, 2-5 cm long, dehiscing by transverse pores. The fruit bears a persistent disc at the apex formed by the union of the stigmas of the flower. These plants are extensively cultivated in Turkey, Yugoslavia, India, Persia and Russia.

The coagulated latex obtained by incising the ripened capsule provides the drug. It contains more than 23 alkaloids most of which are combined with meconic acid ($C_5 HO_2 (OH) : (COOH)_2 3H_2O$). The drug contains sugars, salts, proteins and colouring materials. Of the principal alkaloids, morphine (4-21%), codeine (0.8-2.5%) and thebaine (0.5-2%) are highly toxic strong bases, possessing a phenanthrene nucleus. Noscapine (narcotine, 4-8%) and papaverine (0.5-2.5%), are benzyl isoquinolines and are feebly basic. The other major alkaloid is protopine, a weak base. Morphine possesses a phenolic and an alcoholic hydroxyl group and so, when acetylated, forms diacetyl morphine or heroin.

The drug is valued mainly for its morphine content. Turkish and Yugoslavian opium contain 10-16% morphine and 0.3-4% codeine, whereas Indian opium contains only 9-12% morphine and a greater proportion of codeine.

Because of the narcotic properties, opium and morphine are used as pain relievers and as hypnotics. Codeine is a milder sedative and finds use in allaying cough. Morphine and codeine decrease metabolism also. Opium is often used in cases of diarrhoea and as a diaphoretic.

Aristolochia indica Linn. (Aristolochiaceae)

Ishwari, Indian Birthwort

A. indica is a twining perennial with greenish white woody stems, wedge-shaped and three-nerved variable leaves. Flowers are irregular (duck-shaped) having a pale green inflated lobed base narrowed into a cylindric tube terminating in a horizontal funnel-shaped purple mouth and tip clothed with purple tinged hairs. Fruits are oblong capsules enclosing many flat ovate winged seeds. This plant is a native of India occurring in the tropical and subtropical regions.

The roots contain volatile oil, fixed oil and alkaloids. The volatile oil (upto 5%) constituents are sesquiterpenoids such as ishwarone, ishwarane, ishwarene, aristolochene, selina-4, 11-diene and ledol. Fixed oil is present in small amounts (upto 1.7%), and consists of palmitic, stearic, lignoceric, cerotic, oleic and linoleic acids. Aristolochic acid (a derived isoquinoline) is the principal alkaloid. Other alkaloids present are *l*-curine (aristolochine), aristolactam, aristolic acid and aristolamide. Also present in the root are a cytotoxic lignan savinin, rutin, allantoin, *p*-coumaric acid, friedelin, cycloeucalenol and β-sitosterol.

Roots possess emmenagogue, antiarthritic and emetic properties and are also used against leprosy. Aristolochic acid is found to be anticancerous but also causes renal damage. Aristolic acid showed antifertility effect in rabbits.

Aristolochic acid is not a true isoquinoline alkaloid, but because it is presumed to be formed due to oxidation of an aporphine nucleus, it is included here.

Hedera nepalensis Koch. (Araliaceae)

Syn. *H. helix* Linn.
H. rhombea S. & Z.

Ivy

Ivy is a hardy climber, climbing adhesively by the rootlets. Leaves are variable, simple linear lanceolate or cordate-ovate or palmately lobed. Flowers are small in panicled umbels, while the fruit is a globose blackberry. Ivy is common from W. Europe to Japan.

The whole plant yields emetine. Leaves contain hederasaponins such as α- and β-hederins and hederasaponins B&C. Sesquiterpenes like β-elemene, elixene and germacrene B as also β-lectins are other components of leaves. Inflorescence contains β-amyrin, β-sitosterol and its glucoside, oleanolic acid and triterpene glycosides nepalin-1 (hederagenin arabinoside), napalin-2 and napalin-3.

The plant, in various combinations, is used against cancer and indurations of different kinds. Its extract is found to be anti-tumour in nature, while the leaves are antirheumatic and antineuralgic. Hederin is found to be acting as a vasoconstrictor and hypotensive. Napalins 1,2 and 3, at concentrations of 0.5%. 0,25% and 0.125%, respectively, immobilize human sperm completely.

Nelumbo nucifera Gaertn (Nymphaeaceae)

Syn. *N. nelumbo* Druce
N. speciosum Willd

Kamal, Lotus

Lotus is a beautiful aquatic herb with stout creeping rhizome, peltate large (upto 90cm or more in dia.) orbicular leaves on long petioles bearing small prickles, large rose/white solitary flowers (sepals and stamens petaloid, sepals, petals and stamens many) and a top-shaped spongy torus containing many uniovulate carpels sunken separately in cavities on the upper side which later mature to ovoid nut-like achenes. This is a native of China, Japan and India.

The dried carpels of lotus contain protein (16%), carbohydrates (66%) and minerals. Leaves, carpels and rhizomes contain three aporphine alkaloids, nuciferine, roemerine and nornuciferine. Another alkaloid nelumbine is isolated from petioles, pedicel and embryo.

Rhizome is given for diarrhoea, dysentery and dyspepsia in children. The paste of its rhizome is applied for ringworm and other skin diseases, and the leaf and flower juice (latex) are used in diarrhoea. The rhizome and carpels are edible.

2.6 Monoterpenoid Alkaloids

About 20 alkaloids, closely related to the monoterpenoid glycoside aucubin, constitute this group. They are distributed in quite unrelated families.

Valeriana officinalis Linn. (Valerianaceae)

Valerian

A native of Europe, valerian is a rhizomatous glabrous perennial. Its rhizome is strong smelling, truncate and sometimes stoloniferous. The stem is erect with pinnatisect leaves with small, white rose coloured fragrant flowers in broadly paniculate corymbs and compressed achenes.

Rhizomes and roots of valerian contain 0.05-0.1% alkaloids, mainly valerine, chatinine and valeriana alkaloid XI. The other important constituents of the rhizome are 0.5-1.0% volatile oil consisting of esters like bornyl isovalerianate, bornyl acetate and bornyl formate, alcohols, terpenes and a sesquiterpene alcohol, valerianol. The oil imparts the characteristic odour to the drug.

Valerian is used as a carminative, antispasmodic, in hysteria and other nervous disorders.

Enicostemma hyssopifolium Verdoon (Gentianaceae)

Syn. *E. littorale* Blue.

Chotta Chirayata/Vavding, Enicostemma

This is an erect glabrous herb, branching from the base with quadrangular stems, sessile, linear-oblong 3-nerved leaves, white small flowers in axillary clusters and an ellipsoid capsule containing many globose seeds. This plant is common in western India.

Aerial parts of Enicostemma yield monoterpene alkaloids such as erythrocentaurine and enicoflavine. Also present are swertiamarin, betuline, swertioside and sylswertin and isoswertisin. The flavonoids present are apigenin, genkwanin, isovetexin, swertisin, and saponarin.

The plant is a restorative, carminative and antidiabetic. Its extract is useful in fever and snake bite.

Tecoma stans Griseb (Bignoniaceae) also possesses monoterpenoid alkaloids like tecomanine, tecomine, tecostanine and tecostidine. The bark of *Tecoma* is used against snake and scorpion bites.

2.7 Phenanthro-indolizidine Alkaloids

Only five alkaloids possessing both phenanthrene and indolizidine nuclei are known at present. These compounds are restricted to a few Asclepiadaceous members like *Tylophora, Cynanchum* and *Vincentoxicum*, and one member of Moraceae, i.e. *Ficus septica* Forst.

Tylophora indica Merrill. (Asclepiadaceae)

Syn. *T. asthmatica* W.& A.

Anantamool, Tylophora

A native of India and Malaya, *Tylophora* is a twining perennial with long fleshy roots and small ovate leaves. The greenish purple flowers are borne on short umbellate cymes. The pair of follicles contains broadly ovate, comose seeds.

All the parts of the plant possess alkaloids. Out of the numerous compounds known, only three are important, i.e. tylophorine, tylocebrine and tylophorinine.

The whole extract of leaves and roots possess powerful vescicant properties—an activity attributed to tylophorine and so used effectively in fighting asthma. Tylophorine has a paralyzing action on the heart muscle, but a stimulating action on the muscles of the blood vessels. All the three principal alkaloids possess anticancerous properties.

2.8 Pyridine-piperidine Alkaloids

The fully characterized alkaloids in this group number about 125. These compounds are widely distributed and reported from quite unrelated groups of plants. They are classified based on the positions of substitutions on the pyridine or piperdine nucleus, e.g. the 1st group has substitutions at position 1 (chavicine), at positions 1 and 3 (arecoline), 1 and 2 (pelletierine), etc. The important alkaloids of this group are arecoline, lobeline, pelletierine, coniine, piperine, nicotine and anabasine.

Areca catechu Linn. (Arecaceae)

Supari, Areca nut

Areca nut is the ripe dried seed of a tall palm, a native of Malaya. It is a slender, monoecious, unbranched palm attaining a height of 25 m or more with a ringed stem and a crown shaft. The petiole of the long pinnate leaves are broad, and stem clasping, and the numerous pinnae are more than 1 m long and long acuminate with upper ones often confluent. The much-branched inflorescences are intrafoliar and bear many staminate flowers at the top and a few pistillate flowers at the base. The drupe is one seeded, orange scarlet in colour with a fibrous pericarp.

The seed of areca nut contains 0.29-0.7% pyridine alkaloids in which arecoline—the principal alkaloid—amounts to 0.1-0.5%. The accompanying alkaloids are arecaine (N-methyl guvacine), arecolidine (methyl ester of arecaine), guvacine (demethyl arecaidine), guvacoline and isoguvacine. The tannin content of the seed varies from 14-26% and consists of gallotannic acid and catechins. Also present in the seed are 13-17% fats similar to coconut oil in composition (consisting of 19.5% lauric acid, 46.2% myristic acid, 12.7%

palmitic acid and small amounts of oleic, linoleic and hexadecanoic acids), 47-80% carbohydrates including saccharose, reducing sugars, galactans and mannans; 5-10% proteins, upto 5% carotenes and traces of sitosterol. The husk of the fruit contains 47-50% cellulose.

Betel (areca) nuts are often chewed with betel pepper (*Piper betle*) leaves and lime for their mild stimulating action. The nuts are used as anthelmintics, especially against tapeworm.

Piper nigrum Linn. (Piperaceae)

Kalimirich, Pepper

Indigenous to India and Indo-Malaya region but cultivated widely in the eastern tropics, this perennial woody climber, the dried unripe fruits of which are known as black pepper, possesses simple, ovate, coriaceous and evergreen leaves. The small flowers are borne in catkins, and the fruits are small, red, one-seeded drupes, about 50 per catkin.

Dried, full-grown but unripe fruits constitute the black pepper of commerce. White pepper is the ripened fruit from which the outer portion of the pericarp has been removed.

Chavicine (an acid amide yielding piperine and chavicinic acid), free piperine ($C_{17}H_{19}O_3N$), piperettine ($C_{19}H_{21}O_3N$) and piperidine are the alkaloids contained in the fruits. The pungency of pepper is due chiefly to chavicine (less than 1%) and a lesser extent to piperine (present in amounts 10 times more than those of chavicine) along with a resin. A volatile oil (1-2.5%) occurs in the fruit, which comprises of α-, and β-pinenes, α-phellandrene, citral, limonene, piperonal, etc. The volatile oil is restricted largely to the pericarp and the pungency to the seed. A fixed oil (<1%) and starch (50%) are also present in the fruits.

Pepper is a popular condiment, and is used medicinally as a stimulant and irritant. It cures cold, cough, and dysentery and increases digestion and appetite. It is also used in cardiac diseases, eczema, diabetes and night-blindness. Pepper exhibits anticonvulsant properties and enhances the bioavailability of antitubercular drugs when administered together. Piperine exhibits CNS depressant, antipyretic, analgesic, anti-inflammatory, antioxidant and hepatoprotective properties.

Piper longum Linn

Pipal, Long Pepper

P. longum of India and *P. retrofractum* Vahl (Chab) of Java, are slender creeping undershrubs with erect and subscandent branches, rooting at nodes. The fruits are small and fused into cylindrical spike-like cones.

The entire unripe dried spikes are used as spices and in medicine. They, in addition to the alkaloids found in *P. nigrum*, contain one more pyridine alkaloid, piperlongumine. The fruits are more aromatic and sweeter than black pepper.

Long pepper is a *rasayana* (tonic) drug, useful in improving memory power and intelligence. It is also an appetizer, aphrodisiac, expectorant and cures dyspepsia, piles, indigestion, anaemia, fever and intestinal worms.

Punica granatum Linn. (Punicaceae)

Anar, Pomegranate

Pomegranate is a small tree, with simple ovate shining leaves, red solitary flowers and a round berry crowned with calyx. The plant is a native of Central Asia, now cultivated everywhere for the very refreshing edible fruits.

The yield of volatile liquid alkaloids from the root and stem barks is about 0.5-1%. The principal alkaloid is pelletierine, accompanied by isopelletierine, methyl isopelletierine and a solid alkaloid, pseudopelletierine. All these alkaloids occur in the form of tannates. The tannin content of the root bark amounts to 22%. The pericarp of the fruit is very astringent and contains about 28% tannins.

The root and stem bark are used as anthelmintic and taenifuge. The pericarp is a reputed anthelmintic for children.

Conium maculatum Linn. (Apiaceae)

Poison Hemlock

Poison hemlock is a much-branched herb with a purplish red speckled stem, pinnately decompound leaves, small while flowers in large, many-rayed umbels and laterally compressed 5-ridged cremocarp devoid of oil tubes. Hemlock is native of Europe, occurring as a weed in America and some parts of Asia.

All parts of this plant contain liquid alkaloids, highest in the fruit. The dry unripe fruits, which form the drug, contain 1-2.5% of the liquid alkaloid coniine, along with other alkaloids like N-methyl coniine, conhydrine, pseudoconhydrine and γ-coniceine. The alkaloid content decreases with the age of the fruit. The seeds also contain coumarins such as bergaptene and xanthotoxin.

The unripe dry fruits—the conium of commerce—are used as antipasmodic, sedative, anodyne, in epilepsy, acute mania, and whooping cough and for treating obstruction of lymphatic system. In homeopathy, it is used to prevent cataract.

Lobelia inflata Linn. (Lobeliaceae)

Lobelia

A plant indigenous to N. America, lobelia is a hairy annual, 1 m in height. The leaves are ovate-lanceolate. The light blue flowers are borne in loose spike-like racemes. The fruit is a capsule enclosed by an inflated calyx.

The plant yields 0.25-0.4% alkaloids, a pungent volatile oil, resins, lipids and gums. Of a total of sixteen alkaloids isolated, lobeline is the major compound. The important minor alkaloids are lobinaline, lobelidine, lobelanine, lobelanidine and isolobenaline. Lobenaline is peculiar in possessing both pyridine and quinoline nuclei.

Lobelia is used as an expectorant in chronic bronchitis and asthma.

Indian Lobelia, *L. nicotinaefolia* Heyne, (Narasala), a substitute of *Lobelia* possesses larger leaves and stem and yields not less than 0.8% alkaloids.

2.9 Quinazoline Alkaloids

Vasicine and related alkaloids form the important members of this group.

Adhatoda zeylanica Medic. (Acanthaceae)

Syn. *A. vasica* Nees

Vasaka, Malabar Nut

Vasaka is a tall perennial shrub with elliptic–lanceolate hairy leaves, large white flowers in large bracteolate spikes and clavate capsules containing globular seeds. It is found naturally or cultivated throughout India.

The major alkaloids of the leaves are quinazoline compounds such as vasicine (peganine, upto 1.1%), vasicinone and vasicinol. More than half of vasicine occurs in the *l*-form and the rest in its recemic form. Vasicinone also occurs in *l*-form but racemises on standing. The other components of leaves are a steroid vasakin, betaine, β-sitosterol and an essential oil (containing limonene). Flowers also are found to contain vasicine, vasicinine, β-sitosterol, β-amyrin, essential oil, luteolin, quercetin and kaempferol. Roots yield vasicine, vasicinol and tritriacontane. Seeds yield upto 25% of a fatty oil rich in oleic and linoleic acids. This oil contains appreciable amounts of behenic, lignoceric and cerotic acids. β-sitosterol also is obtained from the seeds.

Vasicine and vasicinone are well-known bronchodilators, respiratory stimulants, hypotensives and cardiac depressants. Vasicine exhibits oxytocic and abortificient activity. The leaves, flowers and fruits are used for treating cold, cough, chronic bronchitis, and as a sedative expectorant, antispasmodic and anthelmintic. The leaf juice is used against diarrhoea, dysentery and glandular tumour and as emmanagogue. The leaves possess insecticidal and weedicidal properties.

2.10 Quinoline Alkaloids

About 100 alkaloids possess the quinoline nucleus. They have a very restricted occurrence and so are reported from *Cinchona* of Rubiaceae and a few members of Rutaceae.

Cinchona spp. (Rubiaceae)

Cinchona

A number of species of *Cinchona* provide important alkaloids of which the major sources are *C. officinalis* Linn. (yellow cinchona), *C. calisaya* Wedd., *C. succirubra* Pac. (red cinchona) and *C. ledgeriana* Moens.

Cinchona officinalis is a native of Peru, now cultivated extensively in the United States, Latin America and India. The plant is a tall handsome tree with an ashy bark. The leaves are simple, acuminate at both the ends. The white flowers occur in terminal corymbose panicles.

Even though alkaloids are distributed throughout, the dried barks of roots and stems are the richest sources. The total number of alkaloids exceeds 25, of which the important ones are quinine, quinidine, cinochonine and cinchonidine. The average yield of alkaloids is about 6-7% of which 60% is quinine, but in red cinchona (*C. succirubra),* cinchonidine is present in larger proportions. All the alkaloids exist as salts of quinic and cinchotannic acids. The amount of quinic acid is about 5-8% in the drug. Cinchotannic acid is a phlobatannin and is oxidized to produce "cinchona red". Another important constituent of the drug is a glycoside quinovin that is present upto 2%.

Galenicals of cinchona are used as bitter tonics and stomachics. Quinine and its derivatives are extensively used to fight malaria. Quinidine finds use for prophylaxis of cardiac arrhythmias and for the treatment of atrial fibrillation.

Remijia pedunculata Fluck. of Rubiaceae, known as cuprea bark, yields alkaloids similar to those of *Cinchona*. The bark yields 2-6% alkaloids of which cupreine is the major compound. On methylation, cupreine produces quinine.

Toddalia asiatica Lam (Rutaceae)

Syn. *T. aculeata* Pers.

Kanchano, Lopeztree

Kanchano is a prickly, rambling shrub with woody roots, longitudinally wrinkled bark, trifoliolate leaves having oblong elliptic glandular leaflets, and small greenish yellow/white flowers in axillary cymes and fleshy globose orange-coloured fruits (about 1cm in diam.) containing many reniform seeds. The whole plant is armed with recurved prickles. A native of Central Asia, this plant is common in the Western Ghats and Orissa.

The root bark contains alkaloids such as chelerythrine (toddaline), dihydrochelerythrine (toddalinine), skimmianine and 7,8-dimethoxy-2',3'-methylene dioxy-1,2-benzo-phenanthridine. Also present are diosmin, toddalolactone, acceleatin, pimpinellin and isopimpinellin. The leaves yield a volatile oil containing linalool as the principal component.

The root bark is diaphoretic, stomachic and antipyretic and is similar to *Cinchona* in its action against malarial parasite. It is also used as a stimulating tonic, carminative and for treating diarrhoea, cough and gonorrhoea.

Glycosmis pentaphylla DC (Rutaceae)

Asvasakhotah, Glycosmis

This is a small tree with alternate imparipinnate aromatic leaves (leaflets 3-5, obovate/oblanceolate-obtuse), small white flowers in axillary panicles and a cream-coloured globose berry containing 1-3 seeds. Asvasakhotah is common in the forests and waste lands of southern-western India.

The root bark contains alkaloids skimmianine, γ-fagarine, dictamine, arborine, carbazole, 3-methoxy carbazole, homoglycosolone, homoglycolone, glycolone, glycozoline and glycozolicine and β-sitosterol. The roots yields carbazoles glycozoline, glycozolidine, dictamine, arborinine and skimmianine, while the leaves contain arborinine, mupamine (alkaloids) and arborinone (triterpene). Flowers are found to contain glycophymine, glycosolone, glycophymoline and an amide glyomide. Vitexin, 7-OMe apigenin, p-hydroxybenzoic acid, vanillic acid, syringic acid and ferulic acid are the other compounds present in leaves (Jash et al., 1992; Umadevi et al., 1990).

This plant is used for curing cough, rheumatism, anaemia and jaundice, while its leaves are useful in fever, liver complaints, and eczema and skin diseases.

2.11 Quinolizidines/Lupinane Alkaloids

More than 90 of the total 125 alkaloids, which possess quinolizidine or a related nucleus, are reported from the order Fabales. The other members are distributed in the Ranunculaceae, Magnoliaceae, Lythraceae and Asteraceae. The only medicinally important alkaloid is sparteine.

Cytisus scoparius Link. (Fabaceae)

Scoparius, Broom Tops

Scoparius is a woody deciduous shrub of 1-2 m high, with a 5-ridged stem and erect slender branches. The lower leaves are petiolate, trifoliate and the upper ones sessile, possessing only one leaflet. The yellow flowers are solitary and the brownish black pod is densely villous at the margins. It is a native of Europe and W. Asia.

The dried tops yield the chief alkaloid sparteine (upto 1.5%), which is often accompanied by sarothamnine and genisteine. The drug also contains trace amounts of epinine (N-methyl-3, 4-dihydroxy phenyl ethyl amine), tyramine, santiaguine (based on two tetrahydroanabasines), anagyrine and *d*-lupinine (quinolizidines) and large quantities of the yellow flavone, scoparin.

Sparteine is employed as an oxytocic.

Boerhavia diffusa Linn. (Nyctaginaceae)

Syn. *B. repens* Linn.

Punarnava, Spreading Hogweed

This is a highly very variable branched woody prostrate herb with a stout woody fusiform rootstock, a purplish stem swollen at the nodes, long-petioled ovate-cordate often sinuous leaves (whitish beneath) purple/white small flowers in small umbels arranged in panicles and oblong 5-ribbed glandular pubescent fruits. Punarnava is a common weed throughout India.

The entire plant, the drug punarnava, contains two quinolizidine alkaloids punarnavine 1 & 2, hypoxanthine arabino furoside, punarnavoside and upto 6% potassium salts. The roots contain β-ecdysone, β-sitosterol, 5,7-dihydroxy-3, 4-dimethoxy-6, 8-dimethyl flavone, rotenones such as boeravinones A, B,C,D,E &F besides a dihydro iso fuenoxanthone, borhavine, lignans liriodendrin and syringaresinol mono-β-D-glucoside and sugars. The leaves are found to contain 4'-OMe kaempferol and phenolic acids such as vanillic, syringic, gentisic and *o*-coumaric acids (Mangalan et al. 1988; Ahmed & Yu, 1992).

Punarnava, more specifically the roots and leaves, is best used for its diuretic and antiinflammatory properties and therefore opted for treating inflammatory renal diseases, nephritic syndrome, oedema and ascites resulting from early cirrhosis of the liver and chronic peritonitis. It is also recommended for abdominal tumours, cancer and as a cardiotonic.

2.12 Sesquiterpene Alkaloids

Celastrus paniculatus Willd (Celastraceae)

Malkangani/Jyotishmati, Intellect Tree

This is a large woody climbing shrub with a yellow corky bark, nearly circular/elliptic leaves, unisexual greenish white/yellowish green flowers in terminal pendulous panicles and globose yellow capsules containing 2-6 ellipsoid wrinkled seeds enveloped by a scarlet aril. Malkangani is found throughout India.

The seeds of this plant, which form the drug, yield a dark brown oil (52%) consisting of palmitic (31%), oleic (22%), linolenic (22%) and linoleic acids (16%) and a non-saponifiable fraction containing three sesquiterpene alkaloids, celapanin, celapanigin and celapagin and 5 polyalcohol esters.

One of the polyalcohol esters is malkagunin, an acetate bezoate of malkaguniol, and others are esters of polyalcohol A, B, C and D with acetic, benzoic, β-furanoic and β-nicotinic acids. Other compounds reported from seeds are 8β-dihydroagarofuran sesquiterpenoids, celastrine, paniculatin, paniculatadiol, β-amyrin and β-sitosterol. The root bark yields phenolic triterpenoids pristimerin, zeylasterone and zeylasteral, quinone-methide and an oil.

The seed is tranquilizing sedative, emetic, diaphoretic, and febrifugal and is used to sharpen the memory. The oil is used for depression, is tranquilizing and acts as a medicine for hysteria. The roots are used for pneumonia. Leaves are emmenagogue and anti-dysenteric.

2.13 Steroidal Alkaloids

A large group of about 250, steroidal alkaloids possess a typical cyclopentanoperhydrophenanthrene nucleus. Many of them occur as esters of acids like acetic, angelic, butyric and vanillic. Most of the solanaceous steroidal alkaloids remain as glycosides and thereby resemble saponins. They have a limited distribution and are practically confined to families Liliaceae, Ranunculace, Buxaceae, Apocynaceae and Solanaceae.

Veratrum viride Aiton (Liliaceae)

Hellebore

This stout rhizomatous perennial, indigenous to the United States, possesses short thick rootstocks and dimorphic leaves, the lower ones being large and oval and upper ones small and narrow. The yellowish green flowers are borne on panicles and the capsular fruit contains numerous seeds.

Rhizome contains a large number of alkaloids (more than 20) which are grouped into two categories: (1) jerveratrum alkaloids, and (2) ceveratrum alkaloids. Jerveratrum alkaloids contain 1-3 oxygen atoms, occurring in the plant both as free alkamines and as glycosides and are similar to saponins in possessing haemolytic properties. Examples are jervine, veratramine (free alkamines), pseudojervine (jervine + glucose) and veratrosine (veratramine + glucose). Ceveratrum alkaloids are highly hydroxylated, containing 6-9 oxygen atoms and occur as free and/or esters with two or more acids. e.g. germine & protoverine (free alkamines), germinitrine (germine + acetic acid + tiglic acid + angelic acid), protoveratrine A (protovertine + 2 × acetic acid + d-α-methyl butyric acid + α-oxy-α-methyl butyric acid).

These alkaloids, in mild doses, are employed as hypotensives, cardiac depressants and sedatives. They also exhibit a strong insecticidal action.

Veratrum album Ait.—The white hellebore also possesses similar alkaloids.

Holarrhena pubescens Wall. (Apocynaceae)

Syn. *H. antidysenterica* DC.

Kurchi, Holarrhena

Kurchi is a small tree with broadly elliptic leaves, white flowers (corona absent) in terminal corymbose cymes and a pair of long cylindric (often dotted with white spots) follicles enclosing numerous linear-oblong seeds tipped with a spreading deciduous corona of brown hairs. This is a native of India.

The bark of this tree contains steroidal alkaloids (upto 2.5%) such as conessine (upto 0.2%), 7-α-hydroxy conessine, holonamine, dihydroisoconessimine, 3-aminoconan-5-ene and kurcholessine. Other alkaloids reported are regholarrhenine A-F, pubescine, norholadiene, pubescimine, kurchinin, kurchinine, etc. The root bark yields holacetine. Leaves contain alkaloids such as kurchiphylline, kurchiphyllamine, kurchaline, holadysine and holadysamine; aminoglycosteroids like holantosines A,B,C & D and holarosine A and aminodeoxyglycosteroids holarosine B,E and F, flavonoids such as kaempferol and quercetin and phenolic acids (*p*-coumaric, vanillic, syringic, α-resorcylic and *p*-hydroxy benzoic acids). The seeds also contain a good amount of conessine, (Bhutani et al. 1990; Siddiqui et al. 1994; Daniel and Sabnis, 1978).

Bark and seeds are used in dropsy and dysentery (conessine is the antidysenteric principle), stomach disorders, abdominal and glandular tumours. Fruits are anticancerous and hypoglycemic and used to regulate menstruation.

Solanum americanum Linn. (Solanaceae)

Syn. *S. incertum* Dunal
S. nigrum Linn.

Kakmachi/Makoi

This is an erect herbaceous annual, with a much-branched slender glabrous stem, ovate/lanceolate leaves, small white flowers in clusters of 3-5 arising from internodal regions, and globose purplish black/red yellow berries containing many discoid seeds enveloped by an acidic pulp. Kakmachi occurs as a weed throughout India.

The whole plant, especially fruits, is used as the drug. All parts of the plant contain steroidal alkaloids of which solanine and solasodine glycosides are the major components (Sharma et al. 1983). Also present are uttroside A & B and uttronin A (spirostanol and furostanol glycosides), and sapogenins such as tigogenin and disogenin (Varshney and Dube, 1970). The leaves yield glycosides of quercetin (Nawwar et al. 1989).

The plant, especially the fruits, is used as an expectorant, antiinflammatory and diuretic. The plant extract exhibits hepatoprotective and antitumour properties.

Solanum surattense Burm. f. (Euphorbiaceae)

Syn. *S. xanthocarpum* Schrad & Wendl.

Kantakari

This is a diffuse herb with a prickly stem, leaves and calyx. The root is cylindrical and the leaves, ovate-oblong and pinnately 7-10 lobed. Flowers are purple in axillary cymes and the fruit is a globular yellow berry containing many smooth reniform seeds. Kantakari is a common weed throughout India.

The whole plant, which forms the drug, contains steroidal alkaloids such as solasodine (major component upto 0.2%), solamargine, β-solamargine and solasonine; sterols like cycloartenol, norcarpesterol, cholesterol and related compounds.

The drug is a well-known expectorant, antipyretic and useful in bronchial asthma and other respiratory diseases. Solasodine exhibits antispermatogenic, hypocholestrolaemic and antiatherosclerotic activities (Kanwar et al. 1990; Dixit et al. 1992).

Wrightia arborea Mabb. (Apocynaceae)

Syn. *W. tomentosa* Roem. & Shult.

Kutaja, Pala Indigo Plant

Kutaja is a small deciduous tree with abundant yellow latex, tomentose elliptic leaves, white fragrant flowers (with an orange corona of numerous fimbriate scales in 2-3 series) in lax terminal cymes and a pair of pendulous cylindrical grooved follicles rough with tubercles, containing slender seeds comose at the base. This tree is common in deciduous forests of India and nearby countries.

The bark of Kutaja contains alkaloids, conessine, holarrhine, kurchicine and conkurchine. Also present are β-sitosterol, lupeol and α-amyrin. The seeds yield an oil similar to that of *Holarrhena*, while the leaves yield quercetin, kaempferol, iso-orientin, procyanidin and phenolic acids such as vanillic, syringic, *p*-OH-benzoic and sinapic acids (Daniel and Sabnis, 1978)

The root bark and seeds, often substituted for *Holarrhena*, are used for chronic diarrhoea and dysentery. Other uses include curing piles, haemorrhage, cardiac diseases, fever and colic. In Siddha system, this drug is used to cure psoriasis and other skin diseases. The leaves, used as a pot-herb, show analgesic, anti-inflammatory and antipyretic properties. The leaves also yield a blue dye used for dyeing clothes.

W. tinctoria R. Br is a related tree used as a substitute for *Holarrhena*. Its bark is poor in alkaloids but contains β-amyrin, lupeol, β-sitosterol, α-amyrin, ursolic acid and oleanolic acid.

The other important plants yielding steroidal alkaloids are: (1) *Funtumia elastica* Linn. of Apocynaceae (major alkaloid is funtumine), and (2) various other species of *Solanum* yielding solasodine and related compounds. The alkaloids of *Solanum* are of great economic importance as starting materials in the synthesis of steroidal hormones. β-solamarine from *S. dulcamera* Linn. is tumour inhibitory and tomatine from *Lycopersicum* sp. is a good antibiotic.

2.14 Tropane Alkaloids

Tropane alkaloids form a small family of about 40 members, of which more than half occurs in plants belonging to Solanaceae, a dozen in Erythroxylaceae and one each in Euphorbiaceae, Convolvulaceae, Lentibulariaceae, Brassicaceae, Rhizophoraceae and Asteraceae. The tropane nucleus is dicyclic and formed by the condensation of ornithine (a pyrrolidine precursor) and three acetate units. The fact that both pyrolidine and piperidine ring systems can be visualized in the tropane skeleton prompted some authors to merge these alkaloids in piperidine or pyrrolidine groups.

Atropa belladonna Linn. (Solanaceae)

Belladonna

The dried leaves and flowering/fruiting tops of *A. belladonna* constitute the drug. This plant is native to Central and S. Europe but is extensively cultivated in the United States, Europe and India. Belladonna is an erect branching leafy perennial about 1 m in height. The leaves are simple, alternate (leaves of upper branches are in pairs owing to the adnation of petiole to the stem), ovate and about 8-15 cm in length. The flowers are blue, purple or red, solitary or in pairs and the fruit is a black, globular, many-seeded berry.

The alkaloid content of the plant reaches the maximum when the fruit setting begins. The leaves yield 0.35-1.00% alkaloids of which *l*-lyoscyamine amounts to 75% and atropine 25% along with a small amount of volatile bases such as pyridine and N-methyl pyrroline. (Hyoscyamine, $C_{17}H_{23}NO_3$, is an ester of tropine- a 3-hydroxy derivative of tropane- with *l*-tropic acid and atropine is the racemized (*dl*) form of hyoscyamine.) The roots also contain a few other alkaloids like apotropine (the anhydride of atropine), and belladonine (stereo-isomer of apotropine), cuscohygrine and scopolamine in traces. Leaves contain a coumarin, β-methyl aesculetin (scopoletin) and calcium oxalate crystals.

Leaves are commonly administered in the form of belladonna tincture as sedatives and to check excessive secretions. The root preparations are used externally to relieve pain. Hyoscyamine and atropine, in pure form, act as anticholinergic. Atropine produces a stimulant action on the central nervous system. The property of atropine and hyoscine (scopolamine) to dilate the pupil of the eye is frequently employed in ophthalmic practices.

Hyoscyamus niger Linn. (Solanaceae)

Khurasaniyajvayan, Henbane

A native of Europe, henbane is now extensively cultivated and frequently encountered as a weed in many parts of the world. It is an annual/biennial, 30-90 cm in height, with fusiform roots. The lower leaves are petiolate but the upper ones are stem clasping and decurrent. The subsessile greenish yellow flowers are grouped in simple, one-sided terminal spikes. A persistent calyx encloses the fruit, a two-celled pyxis, containing numerous reniform seeds.

The dried leaves, collected at the time of flowering, contain 0.05 to 0.15% alkaloids, 75% of which is hyoscyamine and the rest, scopolamine. Compared to the leaf blade, the petiole yields more alkaloids.

The action of henbane is similar to belladonna but weaker. It is often used to relieve spasms of the urinary tract and added to strong purgatives to prevent griping.

The commercial source of hyoscyamine is *Hyoscyamus muticus* L., Egyptian henbane, which yields about 1.5% total alkaloids, is composed largely of hyoscyamine.

Datura stamonium Linn. (Solanaceae)

Dhatura, Stramonium/Thorn Apple

This Asian plant is an annual, 1-2 m in height. The stem shows dichasial branching with alternate slightly pinnatified leaves. The large white flowers are solitary in the axils of leaves. The quadrilocular erect capsule is very prickly and contains numerous brown reniform seeds. Stramonium is often cultivated in the US and Europe, but occurs as a weed all over the world.

The plant is considered as one of the most poisonous ones. The drug stramonium constitutes the dried leaves and flowering tops. The total alkaloid content of the drug is between 0.25 and 0.5%, of which the major alkaloids hyoscyamine and hyoscine occur in the ratio 2:1. But in young shoots, hyoscine has been found in higher concentration. The seeds yield about 0.2% alkaloids and 15-30% fixed oil. The stramonium root contains ditigloyl esters of 3, 6-dihydroxy tropane and 3, 6, 7-trihydroxytrapanes and some alkamines along with hyoscyamine and hyoscine. Leaves contain flavonoids such as chrysin, liquiritigenin, naringenin, kaemferol, quercetin and a withanolide, withastramonolide (Glotter, 1989; Lakshmi and Krishnamoorthy, 1991).

Stramonium is often used as a substitute for belladonna. The vapour produced at the time of burning of the powdered seed relieves asthma. On ingestion, the drug produces narcotic effects. Both hyoscine and hyosyamine are anticholinergic. Scopolamine causes mydriasis and decreases the action of secretary glands like salivary glands, pancreas and lachrymal glands.

Datura metel Linn.

Kala Dhatura, Indian Dhatura

This plant differs from *D.stramonium* in having purple or bluish flowers.

Leaves contain 0.55% alkaloids, scopolamine being the major one. The alkaloid content of the stem is 0.4% and that of seeds, 0.19%. Other than alkaloids, fresh leaves are found to contain withanolides such as withametelin (daturilin), withametelin B, secowithametelin, isowithametelin, datumetine, datumetixone, daturilinol and daturametelins A-H and G-Ac (Mahmood et al. 1988; Kundu et al. 1989).

D. metel has the same uses as that of *D.stramonium.*

Duboisia myoporoides R. Br. & *D. leichhardtii* F. Muell. (Solanaceae)

Corkwood Tree

D. myoporoides is a small tree with ovate-oblong leaves, white/pale lilac flowers in pyramidal/corymbose panicles and globose berries. This and *D. leichhardtii,* both endemic to Australia, are some of the chief sources of atropine and hyoscine. Apart from these two major compounds, these plants possess other alkaloids like nor-hyoscyamine, tigloidine, valeroidine, poroidine, isoporoidine, valtropine, tigloytropeine, lutropine and apohyoscine. Some other species of *Duboisia,* i.e. *D. hopwoodi* F. Muell. produce nicotine and nor-nicotine in the leaves and tropane alkaloids in the roots. The total alkaloid content and the proportion of individual alkaloids vary with chemotypes and ecological factors. The maximum content of alkaloids reported is 5.0%.

Scopolia carniolica Jacq. (Solanaceae)

Scopolia

Scopolia, a native of Europe, resembles belladonna but is smaller in size. It has a black, slender rhizome and translucent, lanceolate leaves. The fruit is a many seeded pyxis.

The leaves and roots of this plant contain hyoscyamine and hyoscine in high concentrations. The minor alkaloids present in these plant parts are cuscohygrine, 3-α-tigloyloxy tropane, pseudotropine and tropine.

Scopolia is used extensively as a source of hyoscyamine. Sometimes, the rhizome is substituted for belladonna roots.

Mandragora officinarum Linn. (Solanaceae)

Luckmuna, Mandrake/May-Apple

This Mediterranean species is a small herb, 30 cm in height, often associated with much superstition because of the branched, fusiform roots often assuming a human-like form. The oval leaves arise from the base; and are about 30 cm long and undulate. The flowers are solitary, large and greenish yellow. The fruit is a globose, many seeded berry.

The leaves and roots yield the alkaloids atropine, hyoscine, mandragorine and nor-hyoscyamine. The drug also contains scopoletin in larger quantities.

Mandrake is used in place of belladonna. As a poultice, this is used for rheumatic pains, ulcers and for skin diseases.

Erythroxylum coca Lam. (Erythroxylaceae)

Coca

The dried leaves of *E. coca*, a South American shrub, yield coca. The plant, attaining a height of 1-2 m, bears ovate-lanceolate leaves with areolate lines, creamy white flowers and an ovoid, red, one-seeded drupe.

The leaves yield about 0.7-1.5% alkaloids. The 14 alkaloids isolated so far are classified into three groups: (1) *Ecgonines* such as cocaine (methyl benzyl ecgonine), cinnamyl cocaine, methyl ecognine, α- and β-truxillines, etc., (2) *tropanes* like tropeine, pseudotropeine, dihydroxy tropeine, tropacocaine and benzyl tropane, and (3) *hygrines* such as hygrine, hygroline and cuscohygrine. Nicotine also is reported from leaves. But it is the ecogonines that are shown to be the stimulating principles of coca. The composition of the alkaloid mixture varies with the varietal differences and change of environment.

Coca leaves are chewed, as an aid to resist mental and physical fatigue and give great stimulation. By chewing these leaves, the native Indians travel great distances without experiencing fatigue and with the meagrest of food rations. The power of coca in enhancing meditation and incantations and producing trance-like states is responsible for its reputation as a "divine" plant. Its continued use leads to physical deterioration due to malnutrition and resulting in death. Coca is also used in the form of drinks. Cocaine is employed as a local anesthetic.

E. truxillense Miq, a related species, is sometimes used in place of *E. coca*. Because of the higher amount of cocaine in it, this plant is the source of commercial cocaine.

Convolvulus arvensis Linn (Convolvulaceae)

Hiranpag, Small Bindweed

C. arvensis is a prostrate trailing herb with a creeping rootstock, the stem twisted upon itself, ovate oblong leaves which are auriculate/hastate at the base and obtuse and apiculate at the apex, pink white long-pedicelled flowers in axils of leaves and globose capsules containing many subtrigonous seeds. This is a common weed throughout the world.

All parts of the plant contain proteoglycans, β-Me aesculetin, alkanes,α-amyrin and other sterols. Roots contain the alkaloid, cuscohygrine.

The proteoglycan mixture is found to inhibit angiogenesis and stimulate immune responses. The plant extract is hypotensive, increases coronary circulation, and is also said to be anticancerous. Its aqueous extract exhibits mascarinic and nicotinic activity.

Apart from the above-mentioned plants, there are a few species which yield typical tropane alkaloids or their derivatives. Most of them are not pharmacologically exploited. A brief listing of them is done here, hoping that it may benefit the workers who intend to study them further.

1. *Datura sanguinea* R. & P. contains 3α-acetoxy-tropane, littorine and 3α-tigloyloxy-6β-acetoxy tropane.
2. *Datura ferox* L. yields 6β-hydroxy hyoscyamine, tigioidine-3α, 6β-ditigloyloxy-7β-hydroxy tropane and 3α-tigloyloxy-3α-tropane.
3. *Bruguiera sexangula* Poir. (Rhizophoraceae) contains in the bark brugine (tropine 1, 2-dithiolane-3-carboxylate).
4. *Cochlearia arctica* Schlecht. (Brassicaceae) possesses cochlearine, an ester of tropine and *m*-hydroxy benzoic acid.
5. *Convolvulus pseudo-cantabrica* Schrenk (Convolvulaceae) seed contains convolvamine, convolvicine, convolvine and convolvidine.
6. *Physochlaina praealta* Miers. (Solanaceae), leaves yield hyoscine.
7. *Lactuca virosa* L. (Lettuce opium, Compositae), hyoscyamine.
8. *Anthocercis littorea* Labill. (Solanaceae) contains littorine, hyoscyamine and meteloidine.
9. *Phyllanthus discoides* Muell. (Euphorbiaceae) yields phyllalbine, an ester of tropine and vanillic acid.

2.15 Tropolone Alkaloids

This is a group of protoalkaloids possessing a tropolone ring, confined to the Liliaceae. About 25 structures are known, of which colchicine is the lone representative possessing significant medicinal importance.

Colchicum autumnale Linn. (Liliaceae)

Colchicum

Colchicum is an autumn-blooming herb with a tunicate bulb, spring leaves, long tubular crocus-like flowers in scapes and 3-celled capsules containing numerous, hard, reddish brown seeds. This native of middle Asia is cultivated in many parts of Europe and Asia.

The colchicum contains several non-basic alkaloids like colchicine, colchiceine and demecolcine. Colchicine, being the principal alkaloid, can be formed upto 0.6% in the bulb and 1.2% in seeds. The corm also contains plenty of starch. Also present in the seed are colchioside, 3-dimethyl colchicine-3-glucoside, a resin, fixed oil and sugars.

Colchicine effects dispersal of tumours and so is used in the treatment of various neoplastic diseases. It is also used to relieve gout and in biological experiments to produce polyploidy or multiplication of chromosomes in the cell.

Gloriosa superba Linn. (Liliaceae)

Bacchanag, Gloriosa

G. superba is a climbing herb with a rootstock of arched solid fleshy cylindric bifurcately branched tubers pointed at both ends, annual stems, arising from the angles of tubers, sessile ovate-lanceolate leaves having a tendril-like spiral tip, large axillary, solitary, showy flowers (petals linear-lanceolate with crisply wavy margins, colour yellow passing through orange and scarlet to crimson) on a pedicel reflexed at the tip and a large linear-oblong coriaceous septicidal capsule containing many subglobose winged seeds. A native of tropical Africa, Bacchanag is now common in India.

The roots of this climber contain upto 0.37% colchicine and seven colchicine derivatives, sitosterol and β-& γ-lumicolchicines. Seeds and flowers also contain colchicine and its derivatives. Flowers, in addition, contain luteolin, N-formyl-de-Ac-colchicine, β-& γ-lumicolchicines, 3-demethyl colchicine and 2-methyl colchicine.

The roots are purgative, cholagogue, anthelmintic and used in rheumatic fever, leprosy, skin diseases, colic and snake bites. Tubers are abortifacient. Various plant parts are used in spleen complaints, tumours, erysipelas, sores and syphilis.

Premna corymbosa Rottl. (Verbenaceae)

Syn *P.serratifolia* Linn.
P. integrifolia Linn.
P. obtusifolia R.Br.

Agnimanthah, Premna

This is a small tree with yellowish lenticellate bark, small broadly elliptic leaves, small greenish white zygomorphic flowers in terminal pubescent paniculate corymbose cymes and a pear-shaped drupe containing 4 oblong seeds. This is common in the west coasts of India, Sri Lanka and Malaysia.

The roots contain alkaloids, premnine, ganiarine and ganikarine, a colouring matter, tannin and a volatile oil.

One of the *dasamulas*, the roots, as well as leaves are used in medicine. This is a good anti-inflammatory drug used for fevers, heart diseases, neurological diseases and rheumatism. The leaves are also used in fevers, colic and flatulence.

Crataeva nurvala Buch-Ham (Capparaceae)

Syn. *C. religiosa* Hook.f

Varuna, Crataeva

This is a tall tree with trifoliolate compound leaves (leaflets elliptic-lanceolate), white flowers in dense terminal corymbs (sepals petaloid and ovary on a gynophore) and a globose berry containing many reniform seeds embedded in the pulp. A native of tropical Africa, it is now planted in many places in India.

The stem bark yields macrocyclic alkaloids such as cadabacine and its diacetate, flavonoids like catechin, its glucoside and epiafzelechin, a glucosinolate glucocapparin and lupeol. The root bark contains lupeol acetate, spinasterol acetate, taraxasterol and 3-epilupeol. The leaves are found to contain isovitexin, myricetin, proanthocyanidins and phenolic acids such as *p*-hydroxybenzoic, vanillic, ferulic and sinapic acids (Daniel and Sabnis, 1977).

The stem bark is an effective remedy for urolithiasis and lowers the levels of liver glycollate oxidase. It is also used as an anti-fertility and anti-inflammatory factor.

Other pharmacologically important alkaloids, not included in any of the above groups and their sources are as follows:

1. *Cephalotaxus harringtonia* C. Koch. (Cephalotaxaceae) produces several alkaloid esters exhibiting substantial anticancer activity; the important one among them is harringtonine.
2. *Maytenus ovatus & M. buchananii* (Celastraceae) yield a new group of alkaloids designated as "maytensinoids". Of these, Maytensine and related esters possess great antileukemic activity.

3

Terpenoids

Compounds with basic skeletons derived from mevalonic acid, or a closely related precursor, are termed terpenoids. They are considered to be built up of isopentane or isoprene units linked together in various ways, with different modes of ring closure, unsaturation and different functional groups. The isoprene residues are usually linked in a 'head to tail' fashion. However, head-to-head and tail-to-tail linkages as also irregular linkages do occur.

Not all terpenoids comprise carbon atoms in multiples of five, as can be expected of them, considering that they are composed of isopentane units. Some compounds have additional or missing carbon atoms, the modifications (probably) being wrought in the course of the long biosynthetic pathways. Sometimes isoprenoid side chains occur attached to non-terpenoid central skeletons.

The classification of terpenoids is present in Table 3.1.

Table 3.1 Classification of Terpenoids

Class	*Number of carbon atoms*	*Number of isoprenes*	*Sources*
Hemiterpenoids (C_5H_8)	5	1	Volatile oils, esters
Monoterpenoids ($C_{10}H_{16}$)	10	2	Volatile oils, glycosides, mixed terpenoids
Sesquiterpenoids ($C_{15}H_{24}$)	15	3	Volatile oils, bitter principles
Diterpenoids ($C_{20}H_{32}$)	20	4	Resins, chlorophyll
Sesterpenoids ($C_{25}H_{40}$)	25	5	Rare (mostly in animals)
Triterpenoids ($C_{30}H_{48}$)	30	6	Resins, waxes, steroids, saponins, cardiac glycosides
Tetraterpenoids ($C_{40}H_{84}$)	40	8	Carotenoids
Polyterpenoids $(C_5H_8)\eta$	α	η	Rubber and gutta

3.1 Monoterpenes

Monoterpenes are colourless, steam distillable, water insoluble liquids with a characteristic aroma, with boiling points ranging from 140°C to 180°C. These C_{10} compounds are formed by the head-to-tail, head-to-head or tail-to-tail condensation of two isoprene residues, and exhibit every possible mode of ring closure, various degrees of unsaturation, and substitution of different functional groups.

In all, 450 monoterpenes have been discovered (Sticher, 1977), and these can be classified as derivatives of 15 common types of basic skeletons and 15 less common types of basic skeletons (Devon and Scott, 1972).

Based on their chemical structures, monoterpenoids are classified into the following three groups:

(a) Normal monoterpenes
(b) Cyclopentanoid monoterpenes
(c) Tropolones.

Normal Monoterpenes

These monoterpenes include all aliphatic and cyclic (having basically a 6-membered ring), steam-distillable monoterpenes. These occur usually in their free state in steam-distillable oils, and possess a distinct aroma. Recently, non-steam distillable monoterpene glycosides, e.g. β-D-glucosides of geraniol, neral, citronellol, thymol and carvacrol, have been found to exist naturally (Skopp and Horster, 1976).

Based on their ring closures, normal monoterpenes are classified as: (i) Acyclic, (ii) Monocyclic, and (iii) Bicyclic. Functionally, these may be hydrocarbons, aldehydes, alcohols, ketones or oxides.

(i) Acyclic Normal Monoterpenes

Majority of the acyclic compounds are aldehydes. The alcohols occur either freely, or as esters.

(ii) Monocyclic Normal Monoterpenes

The *p*-menthane skeleton appears to be the basis of a majority of monocyclic monoterpenes. The members differ in the number and positions of double bonds and functional groups. Geometrical isomerism is common, resulting from the formation of asymmetrical carbon atoms due to attachment of various functional groups.

(iii) Bicyclic Normal Terpenoids

The bicyclic forms can be visualized as derived from the monocyclic forms by additional ring closures. These are of 7 main structural types.

Normal monoterpenes are employed pharmacologically as skin stimulants, sedatives, expectorants, antiseptics, etc., and as a flavouring for foods and medicines with disagreeable tastes. A few are insecticidal, and hence used as pesticides.

Iridoids (Cyclopentanoid monoterpenes)

This recently recognized group of monoterpenes, is characterized by a cyclopentanopyran ring nucleus. Most of them occur as β-D glucosides (loganin, asperuloside, etc.), but those of the nepetalactone type are without a sugar moiety and are volatile, occurring in essential oils. Secoiridoids are compounds arising from loganin by the cleavage of the cyclopentane ring (gentiopicroside, oleuropin and swertiamarin).

Both iridoids and secoiridoids are bitter in nature and contribute much to the bitter taste of many medicines. Both these groups of compounds are found to exhibit a wide range of medicinal properties. For example, all of them are antimicrobial in nature; allamandin from *Allamanda cathartica* is antileukemic; oleuropin from olive as well as valeopicrates (a group of non-glycosidic compounds) from *Valeriana* spp. are found to be hypotensive and sedative; herpagide from *Herpagophytum procumbens* is analgesic and geniposide from *Gardenia jasminoides* is found to be a laxative. Amarogentin and amaroswerin, two secoiridoid glycosides from gentian root, are among the most bitter substances.

Tropolones

These compounds do not contribute to the flavor, but have been included here owing to their terpenoid nature. Tropolones have a seven-membered ring with a double bond system conjugated with a keto group and a hydroxyl group, as their basic structure. They are restricted in distribution, identified as yet only in certain fungi and conifers, e.g. γ-thujaplicene in *Thuja*. Tropolones resemble phenols structurally and in being strong fungicides. They occur freely or are linked with other compounds. Tropolones are reviewed by Erdtmann (1955) and Pauson (1955).

The C_{10} monoterpenes are also present in combination with a large number of compounds. These compounds, sometimes named as *mixed terpenoids*, are alkaloids (many of the indole alkaloids such as iboga, aspidosperma and corynanthe types); phenols (cannabinoids); ketoalcohols (pyrethrins); flavonoids (rotenone, calophyllolide, etc.) or coumarins (marmelosin).

3.2 Sesquiterpenes

More than 1,200 sesquiterpenes are known today. Their structures are based on 30 main skeletal structures, of which there are around 700 compounds; and 70 less common skeletal structures, of which there are around 500 compounds.

Sesquiterpenes are composed of three isoprene units, i.e. are C_{15} compounds; and occur in steam-distillable volatile oils, and in the bitter principles of many plants. Sesquiterpenes contribute to the flavour, and are classified into the following three groups: (a) Acyclic, (b) Monocyclic, and (c) Bicyclic.

Farnesol is the most important acyclic sesquiterpene widely distributed amongst plants. Its pyrophosphate is a key intermediate in terpenoid biosynthesis (Robinson, 1975). Abscissic acid, a growth regulator, is a sesquiterpene carboxylic acid, related structurally to the carotenoid violaxanthin.

Bicyclic sesquiterpenes are based structurally on either the naphthalenic or the azulenic (one cyclopentane and one cycloheptane rings) type of basic skeleton.

Sesquiterpenes are known to exhibit a number of biological and pharmaceutical properties. Guaiazulene and chamazulenes are antiinflammatory. Bisabolol is antiphlogistic and spasmolytic as also petasin and isopetasin (from *Petasites hybridus*). Illudin M. and S., tumigillin, etc., are antibiotics. Many of the sesquiterpenes are bitter to taste and are the constituents of many bitter medicines. Asteraceae are particularly rich in these compounds. A large number of sesquiterpenes are found to exhibit antitumour activities. They are germacranolides, guaianolides and pseudoguaianolides, elemanolides and related compounds. Germacronolides form the largest group here. The important members of this group are the following.

1. Elephantin, elephantopin, molephantin, phantomolin, molephantinin and deoxyelephantopin from various species of *Elephantopus*
2. Onopordopicrin from *Onopordon acanthium*
3. Vernomygdin from *Vernonia amygdalina*
4. Eupatoripicrin, eupacunin, eupacunoxin, eupatocunin, eupatocunoxin, eupatolide (from various species of *Eupatorium*)
5. Liatrin from *Liatris*
6. Ovatifolin, erioflorin from *Podanthus*
7. Alatolide from *Jurinea*
8. Ursiniolide from *Ursinia*
9. Cnicin from *Cnicus*
10. Tagitinin from *Tithonia*
11. Costunolide, epitulipinolide and lipiferolide (*Liriodendron*)

Guaianolides and pseudogauianolides are also equally large groups containing the following members.

1. Gaillardin from *Gaillardia*
2. Eupatorin, eupachlorin, eupatoroxin from *Eupatorium*
3. Damsin, damsinic acid from *Ambrosia*
4. Cyanaropicrin, grosheimin from *Cynara*
5. Zaluzanin C from *Zaluzania*
6. Angustibalin, helenalin from *Balduina*
7. Autumnolide, microlenin, microlenenin A from *Halenium*
8. Multiradiatin, radiatin, pleniradin, fastigilin from *Baileya*
9. Trilobolide (*Laser*)

Elemanolides and related compounds are vernolepin, vernomenin, vernodalin (*Vernonia*), arnicolide (*Arnica*) and eremantholide A (*Eremanthus*). Expect for *Liriodendron* (Magnoliaceae) and *Laser* (Apiaceae), all other plants containing sesquiterpene lactones belong to the Asteraceae.

3.3 Volatile Oils

Both monoterpenes and sesquiterpenes are the components of volatile oils. Volatile oils are volatile aromatic liquids present in special ducts or oil glands. The aroma is due to a principal component or a group of components often modified by the other compounds present. Monoterpenes form the major components of a large number of oils; though sesquiterpenes are present in many oils in varying amounts, they predominate.

Various essential oils and their constituents are generally used against infections, particularly those of the bronchial and urinary tracts and in preventing sepsis of burns and wounds. The antiseptic activity of certain monoterpenes like thymol is about 20 times more than phenol. Thymol and carvacrol are used in mouthwashes and other monoterpenes like eugenol, menthol are employed in toothpastes for their antiseptic and rubefacient actions that are beneficial to gums. Some, like ascaridole, are important anthelmintics. Due to the irritant properties, some are used as counter irritants and as rubefacients in the form of embrocations and liniments. They produce an initial feeling of warmth and smarting, which is often followed by a mild local anesthesia making them useful in antipruritic preparations. Such preparations alleviate rheumatic pain and neuralgia and can be administered in common colds and bronchitis. Due to their mild irritation of the bronchial glands, they are also used as inhalants with expectorant and cough stimulant properties. Volatile fractions of poppy seeds, sacred basil and cumin seeds, on interacting with glutathione-S-transferase, transform into anti cancer agents and inhibit cancer development (Aruna and Sivaramakrishnan, 1996).

Certain essential oils contain monoterpenes, which act on the central nervous system and thus can be exhibit central stimulating, central sedative and narcotic effects. Oils of valerian, calamus, melissa and lavender exhibit a sedative effect. The sedative property is attributed to citronellal, geraniol, citronellol and citral, the first showing the greatest activity. Camphor is said to have a central stimulating effect. The carminative property exhibited by certain oils is due to local irritation (by which the enzymes are released) and due to the spasmolytic effect.

Sulphur-containing oils, containing sulphides and isothiocyanates, being volatile, are included here.

3.4 Aromatherapy

Aromatherapy is a relatively new term for an ancient practice of using fragrant essential oils to treat a wide range of physical and emotional ailments. A French chemist Rene-Maurice Gattefose coined the term aromatherapy to describe the medicinal uses of essential oils. The oils are directly applied to the affected parts like wounds and burns or massaged into the skin so that the oil reaches the bloodstream or is inhaled into the lungs and the effect of the aroma reaches the mind.

Since the monoterpenes are small molecules and are lipidic, they are rapidly absorbed by the skin. This property is used for chest rubs or abdominal rubs. The antimicrobial, expectorant and local irritant properties of these compounds are fairly known and, therefore, they are often used thus by many people.

Today, aromatherapy is more concerned with the physiological and emotional effects of the fragrance of essential oils than with their antimicrobial and healing qualities. How scents can affect moods, one's state of mind and health by sending messages directly to the brain are subject to an increasing number of research studies.

The part of the brain that receives and processes smells is known as the **limbic system.** This is the oldest part of the brain, the core around which higher brain functions are located. It includes the primary control systems for most of our body functions, including hormone secretion and the autonomic nervous system. Thus, inhaling a fragrance can stimulate the release of hormones, which cause euphoria, pain relief, sexual response, stimulation, control anger or fear, induce sleep or reduce stress.

It is found out that fragrance can affect both mood and behaviour both at home and workplace. People are found to cooperate more when placed in a room with a pleasant smell. The aroma of oils of rosemary, cinnamon, basil, jasmine, basil, black pepper, etc., are found to cause brain-wave patterns much like those caused by caffeine. Researchers in Japan measured the brain waves of women who were receiving facials scented with orange, tangerine and peach designed to lift their spirits. They found that these patterns are similar to those achieved during deep relaxation or meditation. It is also found that the women's blood pressure also lowered. The oils of valerian, neroli, nutmeg and mace cause a lowering of blood pressure.

The fragrances of chamomile, marjoram, sandalwood and bergamot are found to sedate the mind and body. International Fragrance and Flavour, a large perfume company, has acquired a patent to produce a product to ease stress in the workplace by using a blend of neroli, valerian and nutmeg. Adding a few drops of these oils to a warm bath also produce similar effects. In an article in the *British Journal of Occupational Therapy*, aromatherapy is described as a way to promote health and well being through massage, inhalation, baths, compresses, creams and lotions. Using fragrance in hospitals is advocated to reduce stress and depression, sedate or invigorate, stimulate sensory awareness and provide pain relief. Memorial Sloan-Kettering Cancer Centre in New York is already using aromatherapy to lower the stress levels not only of patients, but also their families and staff. (Keville and Green, 1995).

Angelica sinensis Diels (Apiaceae)

Dong quai, Female Ginseng

Dong quai is an aromatic perennial herb with a striate stem, tripinnate leaves, small white flowers in a terminal umbel and a compressed cremocarp. It is a native of China.

The roots yield an essential oil containing ligustilide, safrole, *p*-cymene, butylidene dinaphthide and sesquiterpenes. Also present are coumarin bergaptene and fatty alcohols.

This drug (roots) is useful in relieving menopausal symptoms and for other menstrual disorders and dysmenorrhoea. It is also a restorative in nature and used in general debility and anaemia.

Melissa officinalis Linn. (Lamiaceae)

Lemon Balm

M. officinalis is a lemon-scented perennial herb reaching a height of 1-1.5 m, with simple opposite ovate serrately toothed, deeply veined leaves and small white flowers. Native of southern Europe and W. Asia, this plant is now widely cultivated.

The whole plant yields a volatile oil, flavonoids, and triterpenes. The volatile oil is about 0.1%, consisting of more than 50 compounds of which the major ones, citronellal, β-caryophyllene, neral, geranial, citronellol and geraniol amount to 70%. Also present in the leaves are caffeic, rosemarinic and ferulic acids.

This herb can calm nerves and is a folk medicine for insomnia. It is also used in fevers, colds and spasms. Recently, this plant was found to possess antiviral and anti herpes properties (Dimitrova et al. 1993; Wohlbling and Leonhardt, 1994).

Hyssopus officinalis Linn. (Lamiaceae)

Hyssop

Hyssop is an aromatic herb with opposite lanceolate leaves, irregular lilac flowers in axillary clusters on long narrow spikes and four nutlets. This is a native of the Mediterranean region.

All parts yield a volatile oil (upto 1%) consisting of cineole, β-pinene and a variety of bicyclic monoterpene derivatives (L-pinocamphene, isopinocamphone, pinocarvone, etc.), diterpenoid phenols such as carnosol and carnosolic acid, depsides of caffeic acid, triterpenoid acids (marrubin, ursolic and oleanolic acids), tannins and flavonoids (diosmin and hesperidin). Also present are hyssopin (a glycoside) and resin.

Hyssop (all parts) is an excellent nerve tonic and helps recovery after an illness. It is useful in colic, colds, flu, coughs and flatulence. Hyssop is an emmenagogue and abortifacient too (to be avoided during pregnancy). The polysaccharide fraction and tannins of hyssop are found to inhibit SF-strain of HIV-I virus.

Tanacetum vulgare Linn. (Asteraceae)

Syn. *Chrysanthemum vulgare* Bernh.

Tansy

Tansy is an aromatic perennial (upto 2 m tall) with pinnate leaves (leaflets lanceolate, pinnately divided), yellow disc florets in heads borne on dense terminal corymbs and small 5-angled achenes. This is a native of Eurasia.

Flowers yield a volatile oil rich in thujone and camphor. Sesquiterpenes are also present in the oil. The flowers yield flavonoids and resin.

Tansy is a toxic plant, carefully used for killing intestinal worms. It is also useful in reducing menopausal problems, and must be avoided during pregnancy.

Acorus calamus Linn. (Araceae)

Vacha, Sweet Flag

Calamus is the sweet and aromatic rhizome of sweet flag, growing in marshy places of Europe, Asia and America. The rhizomes are furrowed longitudinally on the upper surface, jointed and spongy. The leaves are long and linear. Flowers are small yellow green, borne in a spadix enclosed by a spathe, and the berries are green, 1-3 seeded with oblong seeds.

All parts of the plant contain an essential oil having asarone (80%), asaraldehyde, acoradin, phenylindane, a phenyl propene [1-(*p*-hydroxy phenol)-1-(O-acetyl) prop-2-ene], calamenol, calamene and sesquiterpene ketones like acorone, calarene and calacone and monoterpenes 1,8-cineole, linalool, α-terpeneol and eugenol methyl ether. The yield of oil is maximum in dried rhizomes (3.5%). In addition, roots yield choline and acoric acid. Acorin, a bitter principle, luteolin glycoside, tannins, mucilage, β-sitosterol, and calcium oxalates are the other constituents reported from the plant.

Vacha is well known for its role in improving memory power (especially for loss of memory) and intellect and is used in treating epilepsy and other mental disorders. It is also employed in chronic diarrhoea, dysentery, abdominal tumours, bronchial catarrh, rheumatism and eczema. The rhizome powder is a well-known insecticide against fleas, bedbugs and used to prevent insect pests in rice and other grains and silverfish in libraries. β-Asarone, a derivative of asarone, is suspected to be carcinogenic.

Alpinia galanga Willd (Zingiberaceae)

Barakulanjar, Greater Galangal

A. galanga is a tall herb with tuberous aromatic reddish brown rootstock, oblong-lanceolate leaves, greenish-white flowers streaked with red, borne on dense-flowered long panicles and orange globose capsules. A native of Indonesia, this plant is cultivated throughout India for its aromatic rhizomes.

Rhizomes contain a volatile oil of a strong and spicy odour, consisting of methylcinnamate (48%), cineole (30%), camphor and pinene. Leaves also yield a volatile oil rich in methyl cinnamate. Seeds are found to contain a volatile oil consisting of l'-acetoxychavicol acetate, l'-acetoxyeugenol acetate, caryophyllene oxide and caryophyllenol and fatty acids.

The dried rhizome is the drug Greater Galangal, a reputed medicine for rheumatism and catarrh. It is also considered to be a tonic, carminative and stimulant and used as an expectorant helping in clearing the voice. The rhizome is effectively used as an insecticide against houseflies. The leaves and flowers are reported to have anti-tubercular properties. l'-Acetoxychavicol acetate and l'-acetoxyeugenol acetate possess anti-ulcer properties.

Alpinia officinarum Hance

Lesser Galangal

A native of China, *A. officinarum* is a perennial herb with thick, creeping reddish-brown rhizomes, lanceolate acuminate ornamental leaves and showy white flowers in racemes.

The constituents of the rhizome, the drug, are a volatile oil and flavonoids. The volatile oil, amounting to 1%, consists of 1,8-cineole, α-pinene, β-pinene, methyl cinnamate, etc. Flavonoids reported are about 14, including quercetin, kaempferol, isorhamnetin, galangin and 3-OMe galangin.

Lesser galangal is used as a spice and carminative.

Amomum subulatum Roxb. (Zingiberaceae)

Bari Elachi, Large Cardamom

A. subulatum is a tall perennial herb reaching upto 2.5 m in height with creeping rhizomes, several erect leafy shoots bearing oblong-lanceolate large leaves and short penduncled glabrous spikes. The capsules are upto 2.5 cm long, obcordate, brownish, containing many flattened seeds. A native of Himalayas, this plant is now cultivated widely in northeast India.

The seeds of this plant yield upto 3% of a volatile oil consisting of cineole (65%), terpeneol, terpenene and pinenes. Other constituents of the seed are protein (6%), starch (43%), anthocyanins such as petunidin; an aurone, subulaurone; chalcone, cardamonin and a flavone, alpinetin. The seeds are used as a substitute for cardamom. They are stimulants and stomachics and are also used in indigestion, abdominal pains and rectal diseases.

Anethum graveolens Linn. (Apiaceae)

Syn. *Peucedanum graveolens* Linn.

European Dill

A. graveolens is an aromatic annual/biennial, upto 1 m in height with glabrous branching stems bearing tripinnatipartite finely dissected leaves. The yellow flowers are borne in compound umbels and the mericarps are oblong, dorsally compressed, having two winged lateral ribs. This is a native of Eurasia.

The fruits of the European dill contain 2.5-4% of volatile oil, 16-20% of fatty oil, and 15% proteins. The major constituents of the volatile oil are carvone (30-60%) and *d*-limonene. Dihydrocarvone, *d*-phellandrene, α-terpenene, carveol and dill-apiole are the remaining components of vol. oil. The fixed oil is rich in petroselinic acid triglycerides. Other compounds isolated from the fruits are myristicin, coumarins (bergaptene, umbelliprenin, umbelliferone, scopoletin and aesculetin), flavonoids (kaempferol, glycoflavone), phenolic acids (caffeic, ferulic, chlorogenic) and steroids (β-sitosterol and γ-sitosterol. The leaves also are found to contain a volatile oil consisting of *d*-phellandrene, α-terpenene, limonene, carvone, dihydrocarvone and eugenol. When the plants age, the oil becomes richer in carvone and limonene content decreases. Vitamin C, vanadium and uranium are the other constituents of the leaves. The inflorescences also yield a volatile oil, phytofluene, β-sitosterol, stigmasterol, scopoletin and umbelliferone.

The volatile oil and its emulsion in water are stomachic, diuretic, anthelmintic and antiflatulent and form an important ingredient in gripe water preparations for children. It has been found that the limonene fraction is active against paramecia and helminthes. The oil is also used in beverages, gargles and toothpastes. Apiole as well as dill are powerful emmenagogues (to be avoided by pregnant women). Dill leaves are used for flavouring curries, beverages and pickles and possess carminative and preservative properties.

Anethum sowa Roxb.

Syn. *A. graveolens* DC,
Peucedanum sowa Roxb.

Sowa, Indian Dill

A. sowa is similar to *A. graveolens* Linn, but the fruits are subelliptical with three longitudinal ridges and vittae having marginal walls appearing irregular.

The fruits yield a volatile oil (1.5 to 3.5%), proteins (20%), carbohydrates (31%), β-sitosterol, piperine and flavonoids (kaempferol and quercetin). The major constituent of oil is carvone (20-70%), with dihydrocarvone, *d*-limonene, *d*-phellandrene, myristicin, apiole and dill-apiole. The tender plant also yields a volatile oil rich in *d*-phellandrene.

Sowa is used as a carminative, antipyretic and anthelmintic and is a component of gripe water. It is also used as a condiment and in confectionery. The seed oil is useful in preventing flatulence in children.

Angelica archangelica Linn. (Apiaceae)

Angelica

A. archangelica, a native of Syria, is a stout aromatic perennial herb, glabrous except for the inflorescence, with bi/tripinnate leaves bearing toothed leaflets, small greenish white flowers in compound umbels and ellipsoid winged mericarps. This herb is cultivated in Europe and W. Asia.

All parts of the plant yield volatile oils. The oil from roots, amounting to 1% dry basis, consists mainly of α-phellandrene. Other constituents of roots are furanocoumarins such as archangelin, prangolarin, oxypeucedanin, ostruthol and osthol, an isocoumarin angelicain and a chromone glycoside. The volatile oil from the fruit (upto 3.8%) contains α-phellandrene, *p*-cymene, azulene and sesquiterpenes. Umbelliprenin, isoimperatorin, bergaptene, prongolarin and ostruthol are the coumarins present in fruits.

The roots of Angelica are aromatic and used as diaphoretics, expectorants, and diuretics and in flatulent colic.

A. glauca Edgew, an allied species, and a native of western Himalayas, yields a volatile oil (from roots) containing sesquiterpenes, α-pinene, α-phellandrene and α-cadinene. The root is used as a spice and in medicine in bilious complaints, infantile atrophy, and menorrhagia as also for treating rinderpest.

Apium graveolens Linn. (Apiaceae)

Celery

This strongly aromatic perennial is a native of temperate Europe. It has an erect grooved jointed stem, pinnate (leaflets ovate, 3-lobed) leaves, white small flowers in short-pedicelled/sessile compound umbels and small suborbicular, slightly bitter schizocarps.

The fruits yield a volatile oil (2-3%), a fatty oil (17%), resin, flavonoids, coumarins and alkaloids. The principal component of the volatile oil is limonene (50%), along with selinene (15%), sedanolide and sedanonic anhydride. The last two compounds are responsible for the characteristic odour of the oil. A number of phthalides are also present in the oil. The fatty oil is rich in petroselinic and oleic acids and contains upto 7% resin acids. The flavonoids present are apigenin, luteolin and chrysoeriol. Rutaretin, apiumetin, bergaptene, psoralen and isopimpinellin are the coumarins located. Glutamine, asparagine, caffeic acid and chlorogenic acid are the other compounds isolated from the fruit. From the water extract of seeds, 5 sesquiterpenoid glycosides (celeriosides A-E) and three phthalide glycosides (celephthalides A-C) along with nor-carotenoid glycosides and a lignan glucoside are isolated (Kitajima et al. 2003). Celery leaves yield protein (6.3%), and an essential oil besides the furocoumarins, bergaptene, xanthotoxin and isopimpinellin.

Celery seeds (fruits) are extensively used as spices and possess stimulative and carminative properties. They are used as nervine sedatives and tonics. A decoction of seeds is effectively used against rheumatism and for problems of the liver and spleen. The fatty oil is also used as an antispasmodic and nerve stimulant. The volatile oil is shown to exhibit tranquilizing and anticonvulsant properties. The leaves are used as salad and used in the cure of vitiligo and psoriasis.

Calendula officinalis Linn. (Asteraceae)

Calendula

Calendula is an erect aromatic herb with angular glandular hairy stems, lower leaves spathulate and upper lanceolate having cordate amplexicaul base, light yellow heterogamous flower heads and boat shaped (upto 1 cm long) achenes. This is a native of southern Europe but cultivated throughout India.

The flower heads yield 0.1 to 0.4% volatile oil, consisting of pedunculatine, α-, and β-ionones, *trans*-caryophyllene oxide, carvone, geranyl lactone, *n*-tricosane, terpene alcohols and lactones. The flowers contain a large number of triterpenes such as calenduline, oleanolic acid-based saponins, α-amyrin, β-amyrin, taraxasterol, ψ-taraxasterol, lupeol, brein, faradiol, arnidiol, erythrodiol, calenduladiol, coflodiol (ursadiol), manilladiol, olean-12-ene-3β, 16β, 28-triol, lup-20 (29)-ene-3β, 16β, 28 triol, tarax-20-ene-3β, 16β, 22α-triol, tarax-20-ene-3β, 16β-30-triol, ursa-12-ene-3b, 16b, 21-triol and loliotide acetate. In addition, heliantriol C, oleanolic aldehyde, oleanolic acid, β-sitosterol, stigmasterol, 28-isofucosterol, campesterol, 24-methylene cholesterol, cholesterol, 4β-methyl stigmasta-7, 24 (28) dien-β-ol and 4β-methyl ergosta-7, 24 (28)-dien-3β-ol are also present. The phenolics present are *p*-hydroxybenzoic, *p*-coumaric, gentisic, vanillic, caffeic, gentisic, vanillic, caffeic, *o*-hydroxyphenylacetic, salicylic, ferulic and protocatechuic acids, scopoletin, umbelliferone and aesculetin. The carotenoids of the petals are phytoene, phytofluene, α-, β-, λ-, and ζ-carotenes, prolycopene and lycopene. Also present in the flowers are a bitter principle, loliolide and water-soluble polysaccharides.

The roots contain 8 saponins, calandulosides A-H, all based on oleanolic acid, while the seeds yield protein (33%), oil (44%), [consisting of calendic (53%) and linoleic (34%) acids as major constituents as well as β-carotene], phytin and haemagglutinins (Varljen et al. 1989).

The ray florets form the drug *Calendula*, which is said to possess tonic, sudorific, febrifuge, carminative, antiemetic and anthelmintic properties. Externally, the paste of flowers is used against malignant ulcers, skin diseases, and frost bite. In the USSR, the florets have been used for abdominal cancer. The extracts (aqueous/alcoholic) are sedative and hypotensive in nature and also used against amenorrhoea and leucorrhoea. The leaves, which are used as pot herbs, are useful for treating varicose veins (externally), suppuration and maxillary diseases. Calenduloside B, extracted from the roots, exhibits anti-ulcer, anti-phlogistic and sedative actions. Polysaccharide fractions of flowers exhibited anti-tumour and in vitro immuno-stimulating activity. The flowering herb is the official "Calendula" in Homeopathy.

Syzygium aromaticum Merr. & Perry (Myrtaceae)

Syn. *Eugenia caryophyllata* Thaub.
E. aromatica Bail

Laving, Clove

Cloves are the unopened flowers of a small, evergreen tree, native of the Molucca islands. Cultivated in many tropical countries, the trees have oblong-ovate leaves. In wild state they produce purplish or crimson flowers in terminal branching racemes, but never reach the flowering state when cultivated. The flower buds are greenish or reddish when fresh and have slightly cylindrical receptacles surrounded at the top by 4 calyx teeth enclosing the plump globular unopened corolla. When dried, the buds are brown and brittle.

Cloves contain 14-20% volatile oil, 10-13% tannins, oleanolic acid, vanillin, and a chromene-eugenin. 70-90% of volatile oil is eugenol. Also present in the oil are minor quantities of eugenol acetate, caryophyllene, methyl-n-amyl ketone and some esters.

Cloves are used for flavouring foods and as a stomachic and carminative in medicine. It is also found to heal stomach ulcers and inhibit carcinogens by inducing the production of detoxifying agents such as glutathione S-transferase (Zheng et al. 1992). Clove oil is antiseptic and antispasmodic and is often an ingredient of toothpastes and mouthwashes. It is a toothache remedy. It also finds great use in medicines, perfumery, and histological work and for commercial production of vanillin.

Clove stems, a commercial product, contain only 4-7% of volatile oil and are less aromatic. They are used to adulterate cloves. The dried fruit or 'mother cloves' contain starch and are only slightly aromatic. They also are a common adulterant to cloves.

Marrubium vulgare Linn. (Lamiaceae)

Hoarhound

The white wooly stems of this perennial herb are upto 1 m in height and bear ovate leaves and white flowers in axillary whorls. Hoarhound is a native of Europe and C. Asia.

The leaves and flower tops contain about 0.4% of marrubin, a cyclic diterpenoid bitter principle, an alkaloid *l*-betonicine and an essential oil.

Hoarhound is used as a flavouring agent, and as an antiseptic and expectorant.

Lavandula officinalis Choix (Lamiaceae)

Lavender

This is an evergreen shrub of about 1 m height, with linear lanceolate leaves and blue flowers in terminal spikes. This plant is indigenous to S. Europe.

Fresh flower spikes and leaves yield 0.8-2% volatile oil, comprising of 30-40% esters, chiefly *l*-linalylacetate; geraniol, linalool, limonene, cineole, esters of butyric and valeric acids, and a sesquiterpene.

Lavender is a flavoring agent and used in perfumery as well as an expectorant.

Rosemarinus officinalis Linn. (Lamiaceae)

Rosemary

Rosemary is a native of the Mediterranean region, cultivated in Europe and the USA. These evergreen shrubs are 1-2m in height with slender ash-coloured branches. The leaves are linear and sessile with grey wooly lower surfaces. The mauve flowers are borne in verticillasters.

Leaves and flower tops yield 1-2% volatile oil consisting of bornyl acetate, borneol (the chief constituents), cineole and minor quantities of camphene and α-pinene. Also present in the leaves are rosemarinic acid and triterpene alcohols like α-, and β-amyrins, betulins and β-sitosterol.

Rosemary is known as the "herb of remembrance" and is used to enhance memory and, therefore, recommended for patients suffering from Alzheimer's disease in aromatherapy. It contains a couple of dozen antioxidants and half-dozen compounds that are reported to prevent breakdown of acetylcholine. This plant is a well-known flavouring agent and widely used in perfumery.

Mentha piperita Linn. (Lamiaceae)

Pudina, Peppermint

Peppermint is a small perennial aromatic herb (with an aroma of pepper) having a running rootstock, opposite oblong lanceolate serrate glabrous leaves and purple flowers in thick terminal spikes. It is a native of Europe now cultivated everywhere.

The whole plant yields a volatile oil (1-3%), flavonoids, phenolics and steroids. The main constituents of volatile oil are menthol (30-35%) and menthone (15-30%). Other components are menthyl acetate, menthofuran, cineole, limonene, neomenthol, isomenthol and a sesquiterpene (viridofloral). The flavonoids present are apigenin and luteolin along with highly oxygenated flavones. Caffeic, chlorogenic and rosemarinic acids are the phenolic acids present, whereas the triterpenes located are α-amyrin, ursolic acid and sitosterol.

Peppermint is a carminative and flavouring agent. It is a spasmolytic, smooth muscle relaxant and anti-inflammatory. It helps in the production of bile and is a diuretic as well as a mild sedative.

The yield and composition of volatile oil are influenced by environmental factors, harvesting time and planting time.

Japanese peppermint *Mentha arvensis* Linn. yields a similar volatile oil rich in menthol and is used likewise.

Mentha spicata Linn. (Lamiaceae)

Spearmint

This native of temperate Europe and Asia has now spread all over the world. The plant is a glabrous annual with leafy stolons. Stems are 30-60 cm in height, erect with ascending branches bearing green sessile lanceolate leaves and flowers in interrupted cylindrical spikes.

The yield of volatile oil from the leaves and flowering tops is about 0.5%. The oil contains 55% *l*-carvone (caraway and dill contain *d*-carvone), limonene, phellandrene and some esters. The leaves also contain resins and tannins.

Spearmint is a flavouring agent used in stews, sauces, beverages, chewing gum, dentifrices, etc.

The oil of *Mentha pulegium* Linn. (European penny royal), an allied plant, contains pulegone as its principal constituent.

Thymus vulgaris Linn. (Lamiaceae)

Thyme

This native of the Mediterranean region is a suberect herb, with sessile, linear to ovate leaves borne on slender woody pubescent branches. The lilac flowers are borne in dense head-like resinous clusters.

Dried leaves and flower tops yield 1-2.5% of a volatile oil, 25-40% of which is constituted by thymol. Minor quantities of carvacrol, *p*-cymene, *l*-borneol, linalool, etc., occur in the oil along with thymol. The leaves also contain gums, resins and tannins.

Thyme is a flavouring agent used in soaps, salads, etc. The oil is an antiseptic and is used in perfumery, toothpastes, as fungicides and in internal medicines.

Gaultheria procumbens Linn. (Ericaceae)

Wintergreen

The plant is an evergreen shrub, with a creeping stem and erect leafy branches. The oval leaves are 3-6 cm long, apiculate, glabrous and shining. These are the source of wintergreen. The white flowers are solitary and the fruits are scarlet in colour. This is a native of North America.

The leaves contain gaultherin, a phenolic glycoside that, in the presence of water, is acted upon by an enzyme gaultherase to produce a volatile oil consisting of methyl salicylate (98%), and a sugar, primeverose. The other constituent of the oil is an ester of enanthic alcohol.

Wintergreen oil is used in medicine and flavoring. However, the chief source of the oil of wintergreen today is the sweet birch, *Betula lenta* Linn. that contains gaultherin in the bark.

Caesulia axillaris Roxb. (Asteraceae)

Bangra, Caesulia

A glabrous decumbent/semierect herb with sessile, semi-amplexicaul, oblong-lanceolate leaves, white/pale blue disc florets clustered in the axils of leaves and dark brown ribbed achenes. Bangra a common weed in rice fields throughout India.

The whole plant from Ujjain (Madhya Pradesh) yielded a volatile oil consisting of γ-asarone (48%), limonene (34%), cineole (4%) and pinene (3%). Another sample from northwest Himalayas is found to contain volatile oil consisting of dilapiole (50%) and methyl heptanol (42%). Catechol, 2,4,5-trimethoxy cinnamaldehyde, 2,4,5-trimethoxy benzaldehyde and asarone are reported from aerial parts. Seeds are found to yield fat (15%) consisting of stearic (32%), linoleic (46%), palmitic (15%) and oleic (6%) acids and protein (5%).

The plant is used for treating baldness and goiter and as a diuretic. The young leaves are used as a pot herb. The volatile oil, when isolated, exhibited anti-fungal and anti-bacterial activities as well as anti-feedant activities against insects such as *Spodoptera litura* and *Achoea janata*.

Carum carvi Linn. (Apiaceae)

Kala Zira, Caraway

Caraway is an aromatic glabrous annual/biennial herb reaching a height of 1m, with thick fusiform roots, grooved hollow branches, pinnately dissected leaves having laciniately cleft leaflets, white flowers in compound umbels and oblong-oval yellowish fruits, mericarp curved, brownish with five prominent ribs. This plant is found wild in north Indian hills, and cultivated in a large scale for its aromatic seeds.

The mericarps ('seeds') contain protein (40%), fixed oil and resin (10%) and a volatile oil (3-8%). The principal constituents of the volatile oil are carvone (55-60%) and limonene. Other monoterpenes present are α-pinene, β-pinene, and *p*-cymene. The fixed oil is rich in oleic (60%), linoleic (20%) and petroselinic (17%) acids. The phenolics reported from the seeds are glycosides of kaempferol, quercetin and isorhamnetin and coumarins such as 5- and 8-methoxy psoralens, umbelliferone, herniarin, and scopoletin.

The whole plant also yields a volatile oil consisting of cadinene (37%), carvone (31%) and dihydrocarveol (10%). The roots contain polyenes, falcarindione, falcarinolone, phytofluene, β-sitosterol, umbelliferone and scopoletin. Leaves contain glycosides of kaempferol and quercetin.

Caraway is a popular spice and condiment. It is prescribed in flatulent colic and stomach problems. Its alcoholic extract is antispasmodic. Caraway roots are edible and leaves are eaten as salads.

Chenopodium ambrosioides Linn. (Chenopodiaceae)

Kshetravastuka, Indian Wormseed

Indian wormseed, a much-branched woody herb with a smell of camphor, having angular-ridged purple-tinged fleshy stems, oblong-ovate leaves, green flowers in axillary/terminal spikes and small reddish brown seeds. A native of tropical America, is now found as a weed in waste places.

The flowering and fruiting shoots yield a volatile oil (0.25-0.45%) containing ascaridole (50%), xylene, *p*-cymene, tetradecanol, caproic and caprylic acids. The leaves contain kaempferol and organic acids and the fruits yield glycosides of kaempferol as well as isorhamnetin, quercetin and 4'-O-demethylarbectorin (Jain et al. 1990).

The entire plant is a tonic, emmenagogue, antispasmodic and useful in treating nervous affections. The dry plant is anthelmintic against hookworms and roundworms. The tender leaves and shoots are used as a vegetable. The powdered seed or its decoction is used as an insecticide.

Chenopodium anthelminticum Linn.

American Wormseed

This plant also yields a volatile oil containing 60-80% ascaridole and the amount of oil also is higher (upto 2%). It is used in a similar capacity to the Indian wormseed.

Cinnamomum aromaticum Nees. (Lauraceae)

Syn. *C. cassia* Blue.

Chinese Cassia

C. aromaticum is an aromatic evergreen tree with a smooth grey bark; oblong, elliptic large leaves (3-nerved from base), small flowers in terminal or axillary panicles and black aromatic fruits containing single seeds. This is a native of China.

The bark of Chinese Cassia is rich in diterpenes such as cinnzeylanin, anhydrocinnzeylanin. cinnzeylanol, anhydrocinnzeylanol, cinnacassiols A,

B, C, D_1, D_4 and E and their glycosides. Other compounds in the bark are cassioside, cinnamoside, lyoniresinol glucoside, syiringaresinol, cinnamic aldehyde, epicatechin, cinnamtannins A_1-A_4, procyanidins and a volatile oil (0.3%). The constituents of volatile oil are cinnamaldehyde (82%), eugenol, cinnamic acid, cinnamyl acetate etc. The leaves yield an essential oil (0.4%) containing 71% cinnamaldehyde and 10% eugenol. The flowers and unripe fruits also yield similar volatile oils.

Chinese cassia is used as diaphoretic, antipyretic, analgesic, purgative, cardiac stimulant, diuretic and sedative. The diterpenes are found to be anti-ulcerogenic. The oil is used as a flavouring agent, and in perfumes.

Cinnamomum camphora Presl.

Kapoor, Camphor

All the parts of this tree contain a volatile oil rich in camphor. Camphor separates out on cooling the steam-distilled oil. Other compounds in the mother liquor are safrole, cineole, borneole, camphene, dipentene and terpeneol. A number of varieties of *C. camphora,* produce a volatile oil rich in compounds like safrole, linalool, cineole, etc.

Camphor is used as an anodyne, rubefacient, in rheumatism, as external application on sprains or as a liniment. When taken internally, it is a carminative, expectorant, reflex stimulant of heart and as a nervous depressant in hysteria, epilepsy, etc.

Cinnamomum tamala Nees. & Eberm.

Tamalpathra, Indian Cassia Lignea

This is a moderate-sized tree with dark brown bark, elliptic/oblong 3-nerved leaves, pale yellow flowers in panicles and ovoid black fruits. Two chemotypes of *C. tamala* occur in India. The eugenol-type is found in north India and cinnamic aldehyde-type in eastern India.

The leaves yield an essential oil (0.3-0.6%). The eugenol type contains eugenol (13%), eugenol acetate (12.5%), cinnamic aldehyde (41%), linalool (15%), β-caryophyllene, benzaldehyde, camphor and cadinene. The cinnamic aldehyde type contains linalool (50%), cinnamic aldehyde (12%), α-and β-pinenes, *p*-cymene, geraniol etc. A sample from Assam showed 80-85% eugenol. The leaves contain 3′-methoxy kaempferol, quercetin, free as well as glycosides. The volatile oil from bark contains cinnamic aldehyde in large amounts (80%).

The leaves are used as spice and in colic, diarrhoea and rheumatism. Leaf powder is said to be hypoglycemic. The bark is carminative and used to cure gonorrhoea.

Coleus ambonicus Lour. (Lamiaceae)

Syn. *C. aromaticus* Benth

Karpurvalli/Patjarchur, Indian Borage

This highly aromatic succulent herb is much branched and bears broadly ovate, crenate-serrate fleshy leaves, small zygomorphic purple flowers in dense whorls at intervals in a racemose axis, and a utricle of 4 orbicular nutlets. A native of Malaya, it is now widely cultivated everywhere.

The leaves of this plant contain a volatile oil, large amounts of exalacetic acid along with cirsimaritin, β-sitosterol and other triterpenoid acids. Flavonoids such as salvigenin, 6-OMe genkwanin, chrysoeriol, luteolin, apigenin, quercetin, eriodictyol and taxifolin are reported from South American varieties. The volatile oil consists of thymol (40%) and carvacrol.

This is a common home remedy for cough, cold and fever in children. It also used for kidney and bladder stones, bone fractures, headache, asthma, bronchitis, indigestion, diarrhoea, dysentery and cholera. It is a good carminative, diuretic and stimulates the function of the liver.

Coriandrum sativum Linn. (Apiaceae)

Dhanya/Kottmir, Coriander

Coriander is a glabrous aromatic annual with a slender branching stem. Basal leaves are entire with crenate margin whereas the leaves from stem are pinnate (leaflets branched dichotomously to linear segments but outline of leaflet remains ovate), leaf base almost sheathing. White flowers are borne in umbels in which the outer flowers have enlarged ray-like petals. The spherical cremocarp contains two hemispherical mericarps united by their margins. Each mericarp has five wavy inconspicuous primary ridges alternating with four prominent secondary ridges. Coriander is a native of the Mediterranean region, now widely cultivated everywhere.

The leaves yield isocoumarins like coriandrones A & B, coriandrin and dihydrocoriandrin and are good sources of vitamin C and carotenoids. Fruits contain fixed oil (15%), volatile oil (upto 1%), tannins and calcium oxalate. 65-70% of the volatile oil is linalool. Other components are α- and β-pinenes, limonene, γ-terpenene, *p*-cymene, borneol, citronellol, camphor, geraniol and geranyl acetate. Heterocyclic components such as pyrazine, pyridine, thiazole, furan and tetrahydrofuran derivatives, isocoumarins, viz., coriandrin, dihydrocoriandrin and coriandrones A-E, flavonoids like 6-methoxyquercetin, sterols, phthalides such as neochidilide and 2-digustilide are the other constituents reported from coriander (Gijibels et al., 1982; Tanaguchi et al., 1996; Lamparsky and Linies, 1988).

Coriander is a well-known spice with stomachic, spasmolytic and carminative properties. The essential oil is antifungal in nature.

Cymbopogon citratus Stapf. (Poaceae)

Lilichai, Lemongrass

This is a tall aromatic grass with linear, distichous leaves, elongate panicles and many sessile spikelets. Lemongrass is found wild or cultivated.

The leaves yield a volatile oil rich (78-82%) in citral, waxes, flavonoids (luteolin, isoorientin and orientin), phenolic acids (chlorogenic, caffeic and *p*-coumaric acids), triterpenoids such as cymbopogonol and cymbopogone. Other constituents of volatile oil are *d*-limonene, citronellol, geraniol and myrcene.

Lemongrass is used in elephantiasis, as CNS depressant, analgesic, antirheumatic, antipyretic and expectorant. Lemongrass oil is anticarcinogenic in the sense that *d*-limonene and geraniol induce activity of glutathione transferase, an enzyme that inhibits cancer development. The oil is also used as a remedy for ringworm. Citral is an insect repellant.

Cymbopogon martinii Wats.

There are two varieties of this plant, *sofia* and *motia*, the former yielding 'ginger grass oil' and the latter 'palmerosa oil'. These oils are used in rheumatism.

Eucalyptus globulus Labill. (Myrtaceae)

Nilgiri, Eucalyptus

E. globulus is a tall aromatic tree with grayish white smooth bark exfoliating occasionally and elliptic lanceolate leaves and greenish yellow flowers. Eucalyptus cultivated in many parts as a social forestry tree.

The leaves yield a volatile oil (up to 1.5%) consisting of cineole (65-75%), caryophyllene, α-pinene, limonene, etc. The bark and wood contain polyphenols. Euglobals (acyl-phloroglucinol-monoterpene/sesquiterpene, 11 in number, such as euglobal 1_{a1},1_{a2},1_b,1_c,11_a, 11_b,11_c,111-1X) form another group of compounds isolated from this plant. Also present in the leaves are acetogenin mevalonates, eucalyptin, quercetin, rutin, chrysin, phloroglucinol, oleanolic acid, ursolic acid and phenolic acids such as caffeic, ferulic, gallic, gentisic and protocatechuic acids (Dayal, 1988).

The essential oil is used as an expectorant and antiseptic. A gum resin from plant is used in diarrhoea. Euglobal III is a granulation-inhibiting agent.

Foeniculum vulgare Mill. (Apiaceae)

Madhurika/Saunf, Fennel

Fennel is an aromatic, glabrous erect perennial reaching a height of 1m with finely dissected leaves and yellow flowers in large umbels. The glabrous fruits bear a stylopod at the apex and each mericarp is linear oblong, slightly curved and possesses five prominent primary ridges. The plant is a native of Mediterranean region, now cultivated all over the world.

The fruits contain fixed oil (12%), proteins, a volatile oil (1-4%) and phenolics. The principal component of volatile oil is a phenolic ether, anethole (60%) followed by the ketone, fenchone (10-30%). Other components of oil are methyl chavicol, limonene, α-pinene etc. The phenolics include furanocoumarins (imperatorin, bergaptene, and xanthotoxol), stilbene trimers (miyabenol C and *cis*-miyabenol C and the glycosides of the latter such as foeniculosides 1-4) and flavonoids. Also present are monoterpene glycosides, viz., foeniculosides V-IX, and other compounds such as zizybeoside I, icaviside A, syringin, sinapyl alcohol, adenosine and anethole glycol.

Fennel is used as a spice and carminative. The fruits are anabolic, estrogenic, analgesic, antipyretic, and antimicrobial agents. It reduces flatulence in infants.

Myristica fragrans Houtt. (Myristicaceae)

Jatiphala, Nutmeg

Nutmeg is the seed of an evergreen dioecious tree, a native of Molucca islands. The leaves are simple, ovate, the yellow flowers are fleshy and aromatic. The golden yellow fruits contain a single seed, covered by a brilliant red branching aril, the 'mace' (*Javantri*) of commerce. The seed has an outer dark brown perisperm and an inner light brown endosperm. The perisperm is infolded and penetrates the endosperm, giving it a marbled appearance.

Nutmeg contains fat (24-40%), occurring as prismatic crystals-nutmeg butter; volatile oil (5-15%), phytostearin, large amounts of starch, protein and amylodextrin. The volatile oil consists of 80% terpenes like pinene, camphene and sabinene, 8% dipentene, 4% myristicin, 2% elemicin and isoelemicin; upto 1% safrole and 1% other constituents like eugenol, methyl eugenol, isoeugenol, etc.

Nutmeg, in doses larger than 10-15 gm (one teaspoonful), is found to produce intoxications and hallucinations-similar to LSD-after a delay of 3-5 hrs in many individuals. This intoxication is often accompanied/followed by certain side effects like headache, absence of salivation, dizziness, etc. The active principles responsible for this are not clearly pinpointed, but are found to be present in the volatile oil fraction of nutmeg. It is believed that both myristicin and elemicin—two major ingredients of aromatic fraction of volatile oil—are involved in this process. The myristicin content varies from 3.9-13% and that of elemicin from 0.02-2.4% and their relative amounts account for the variations in the psychotomimetic properties of different samples.

Nutmeg is a carminative efficiently used against indigestion and other stomach problems of children.

Ocimum tenuiflorum Linn. (Lamiaceae)

Syn. *O. sanctum* Linn.

Tulsi, Basil

This is a perennial herb with a brownish pink stem, elliptic-oblong pubescent (minutely gland-dotted) leaves on a hairy slender petiole, whitish pink flowers on terminal racemes and a carcerule fruit enclosing 4 nutlets. Basil is a native of India, cultivated as a sacred plant.

The leaves yield a volatile oil (upto 0.8%), triterpenes, flavonoids and fatty acid esters. Eugenol (content of which is maximum in spring, upto 20%) and caryophyllene (38%) are the principal components of vol. oil. Other constituents are bornyl acetate, β-elemene, methyleugenol, neral, β-pinene, camphene and α-pinene. Ursolic acid, campesterol, cholesterol and stigmasterol are the triterpenoids present.

Tulsi is known to possess hypoglycaemic, immunomodulatory, antistress, analgesic, antipyretic, antiinflammatory, antiulcerogenic, antihypertensive, CNS depressant, radioprotective and antitumour properties.

Curcuma zedoaria Roscoe. (Zingiberaceae)

Syn. *Ammomum zedoria* Christm.
Curcuma zerumbet Roxb.

Karcurah, Zedoary

Zedoary is a large perennial with rootstock having palmately branched sessile cylindric oblong annulate yellowish white tubers and a camphoraceous odour, a pseudostem of sheathing leaf bases, long-petioled oblong leaves clouded with purple down the middle, yellow flowers in axils of pouched bracts in dense strobiliform spikes, (stamen one, staminodes petaloid) and trigonous globose capsule containing ellipsoid seeds having a white lacerate aril. This plant is cultivated and also grows in the wild in most parts of India.

Rhizomes are found to contain sesquiterpenes such as curcumol, curcolene, procurcumenol, isocurcumenol, furanodiene, curcumadiol, dehydrocurdione, zederone, etc. Other compounds present are ethyl-*p*-methyl cinnamate and curcuminoids. Also present is a volatile oil containing sesquiterpenes zedoarone and curzerenone.

The rhizomes are used mainly for cough, bronchitis, asthma, skin diseases, wounds and splenic disorders. It is also a carminative.

Piper betel Linn. (Piperaceae)

Tambulah/Pan, Betel

This plant, a native of Malaysia, is a perennial dioecious vine with semi-woody stems climbing by means of adventitious roots, broadly ovate-cordate leaves, dense cylindrical male spikes and long pendulous female spikes.

The leaves, used for chewing, contain up to 2.6% of bright yellow aromatic oil which is extracted by steam distillation. Chavibetol (an isomer of eugenol) forms the characteristic component of betel oil. But some of the Indian samples contain eugenol in larger quantities (25-40%). Other constituents of oil are chavicol, eugenol methyl ether, cineole, caryophyllene and cadinene.

Betel is a carminative, stimulant, anthelmintic, expectorant and antiseptic and cures dyspepsia, fever, flatulence and filariasis. Oil of betel is used in the treatment of respiratory troubles, as an antispasmodic and a carminative. It is antibacterial in nature and is an effective antioxidant. Roots cause sterility in women.

Ruta graveolens Linn. (Rutaceae)

Garden Rue

A strongly odoriferous glabrous herb reaching a height of about 1m with 2-3 pinnate leaves having oblong/spathulate strongly aromatic leaflets covered with a bloom, small yellow flowers in corymbs and small capsules. A native of Mediterranean region, it is now cultivated widely.

The herb contains a volatile oil, alkaloids, flavonoids and coumarins. The volatile oil, rue oil, is strongly odoriferous, pungent and contains methylnonyl ketone (80-90%), methyl heptyl ketone, α-pinene, limonene, cineole, etc. Rutin amounts to 2% of the plant. Imperatorin, isoimperatorin, xanthotoxin, bergaptene and psoralen are the coumarins present. The alkaloids obtained from the stem and leaves are skimmianine, graveolinine and kokusaginine, whereas the roots yield dictamine and γ-fragarine.

Rue oil is anthelmintic, antispasmodic, antiepileptic, rubefacient and emmenagogue. The herb is diuretic, emmenagogue and antispasmodic. As a poultice, the herb is applied to rheumatic joints. Coumarins are responsible for the antispasmolytic activity.

Hyptis suaveolens Poit. (Lamiaceae)

Gangatulsi, Hyptis

This is a rigid erect aromatic annual with petiolate ovate (lower cordate) serrate leaves and blue flowers in axillary cymes and a carcerule containing 4 elongate black nuts. A native of tropical america, Hyptis is grown in gardens or found as an escape.

Leaves, twigs and flowers contain a volatile oil consisting of sabinene, α-pinene, limonene, β-caryophyllene, 1,8-cineole, spathulenol, caryophyllene and bicyclogermacrene. Aerial parts contain a triterpenoid, hyptadienic acid.

This plant is a stimulant, carminative, antiseptic, sudorific and galactogogue. Infusion of the plant is used in catarrhal conditions, affections of the uterus, diabetes and cancer.

Iridoids

Barleria prionitis Linn. (Acanthaceae)

Sahacharah, Barleria

This is a spiny much-branched shrub with sharp long axillary spines, elliptic/obovate long-petioled leaves, yellow flowers (solitary in lower axils, spicate above) having spiny tipped bracts, and ovoid beaked capsules containing 2 compressed hairy seeds. *B. prionitis* is very common in India and also grown as a hedge plant.

The leaves and stems are found to contain five iridoid glucosides, barlerin, acetyl barlerin, shanzhiside methylester, 5,6-β-epoxy, 7β-hydroxy, 8β-methyl-1β-rhamnosidal iridoid, gentioside, 4-carbomethoxy, 5,6-dehydro, 7,8-dihydroxy, 8-methyl-1β-D-glucopyranosidal iridoid (Taneja & Tiwari, 1975) as well as β-sitosterol. The leaves yield a 6-hydroxy flavone, scutellarein and phenolic acids, *p*-hydroxy benzoic, vanillic, syringic, and melilotic acids (Daniel & Sabnis, 1987). Flowers contain scutellarein glycoside.

Sahacharah is widely used against neurological disorders like paraplegia, sciatica, etc. The plant is antiseptic, diuretic and applied over boils and glandular swellings. The bark is used against whooping cough and as a diaphoretic and expectorant.

Barleria cristata Linn.

Raktajhinti, Sahacharah

B. cristata is an erect hirsute undershrub with long-petioled lanceolate leaves, purple blue/pink flowers having serrate spiny bracts/bracteoles, and 4-seeded oblong capsules. This plant is found throughout India.

The roots of *B. cristata* contain iridoids and anthraquinones barlacristone and cristabarlone. Leaves yield flavones, luteolin and 7-OMe luteolin and phenolic acids, p-hydroxy benzoic, vanillic, syringic and *p*-coumaric acids. Flowers are found to contain β-sitosterol, quercetin, apigenin, naringenin and malvidin.

This plant is used as sahacharah in many places in India.

Barleria strigosa Willd.

Sahacharah

This is another plant used as sahacharah. *B. strigosa* is a tall shrub with ovate decurrent leaves and blue flowers in dense spikes. The whole plant is found to contain β- and λ-sitosterol. Leaves yield apigenin, vanillic acid, *p*-hydroxy benzoic acid and *p*-coumaric acid (Daniel & Sabnis, 1987).

B.courtallica Nees is another plant used in Kerala as a substitute of B. *prionitis*.

Vitex negundo Linn. (Verbenaceae)

Nirgundi, Vitex

Nirgundi is a small tree with grey pubescent branches, 3-5 foliolate leaves (leaflets lanceolate, white tomentose beneath, petioled, lateral leaflets small, lanceolate and terminal leaflets larger), small blue bilabiate flowers (stamens 4-didynamous) and ovoid black drupes containing 4 obovate seeds. This tree is common in India, Afghanistan, Philippines and Ceylon.

The leaves contain, aucubin agnuside; alkaloids nishindine ($C_{15}H_{21}ON$) and hydrocotylene($C_{22}H_{33}O_8N$), glycoflavonoids orientin and isoorientin and 5-hydroxy, 3,6,7,3′,4′-penta methoxy flavone.

The leaves, in the form of a paste, are used for inflammatory swellings of the joints formed due to rheumatism, hydrocele and spleenic enlargement. They are also used in skin diseases, nervous disorders, leprosy and as a vermifuge. Oil prepared with leaves is useful for growth of hair and increases the functions of brain. Roots also are useful in rheumatism, dyspepsia, piles and as anthelmintics.

Vitex trifolia Linn, a related species with 3-foliolate, sessile, obovate leaflets and blue flowers are used as a substitute for *V. negundo.*

Picrorhiza kurroa Linn. (Scrophulariaceae)

Katuka, Picrorhiza

Katuka is a hairy shrub, native to the Himalayas, with perennial bitter woody rootstock clothed with persistent leaf bases. The subradical leaves are ovate with winged leaf bases and the white/blue flowers are borne on a stout leafy scape. The ovoid capsule contains a number of seeds, each enclosed in a large bladdery, loose and hyaline reticulate testa.

The major constituents are two iridoid glycosides, picroside 1 and kutkoside. Also present are other iridoids picroside III, veronicoside, minecoside; phenol glycosides picein and androsin, a number of cucurbitacin glycosides and 4-hydroxy-3-methoxy acetophenone (Stuppner and Wagner, 1989).

Kutkin as well as the whole extract show hepatoprotective activity. It is also an immunostimulant of both cell-mediated and humoural immunity (Ansari *et al.* 1988b).

Gentiana lutea Linn. (Gentianaceae)

Gentian

The fermented rhizomes and roots of *G. lutea* constitute this drug. It is a perennial herb, a native of America and Europe, with very long roots (often more than 1m long), big oval rosette leaves, 3-10 yellow long-pedicelled flowers in axillary clusters and capsular fruits.

The rhizome and roots yield more than 2% of a bitter iridoid principle known as gentiopicrin (gentiopicroside, gentiamarin) which, on hydrolysis, yields a lactone gentiogenin and glucose, along with related glycosides amaropanin, amarogentin and amaroswerin. The other components of the drug are 0.6-1% of monoterpenoid alkaloids gentianine and gentialutine, swertiamarin and sweroside, flavones, glycoflavones isoorientin and isovitexin, β-amyrenol, xanthones, tannins, gentianose (a trisaccharide) and gentiobiose (a disaccharide). In addition, roots contain xanthones mangiferin, genistein, isogenistin, gentioside, 1-hydroxy-3, 7-dimethoxy xanthone, 7-hydroxy, 3-methoxy xanthone and 1-hydroxy,3-methoxy xanthone glycosides (Hayashi and Yamagishi 1988).

Gentian is extensively used as a bitter tonic. The root infusion is used for lack of appetite and stomach disorders and as a weak febrifuge.

Sesquiterpenes

Chrysanthemum parthenium Pers. (Asteraceae)

Syn. *Pyrethrum parthenium* Sm.
Tanacetum parthenium Pers.

Featherfew, Feverfew

Feverfew is a small herb with furrowed hairy stem, bipinnatifid leaves having serrated margins and numerous small yellow heads with outer white rays. It is a native of Europe and America.

The principal component of feverfew is a sesquiterpene lactone, parthenolide. Other sesquiterpenes present are santamarine and esters of parthenolide such as reynosin, artemorin, hydroxy costunolide, 8-hydroxyestatiatin, etc. Other constituents of feverfew are volatile oil (containing pinene and its derivatives along with bornylacetate, costic acid, etc.), acetylenes and flavonoid glycosides.

The entire plant is used as a carminative tonic, emmenagogue and employed in hysterical complaints, nervousness and lowness of spirits. It is also recommended for the prevention of migraine and as an antiinflammatory in rheumatoid arthritis. Feverfew is reported to inhibit synthesis of leucotrienes, prostaglandins and thromboxanes, the compounds causing inflammations. In many places, this is a primary remedy in the treatment of migraine.

Nardostachys jatamansi DC. (Valerianaceae)

Jatamansi

A native of alpine Himalayas, jatamansi is an erect, perennial herb with a long woody stout rootstock, often covered by a dense mat of reddish brown fibres formed from the petiole of withered leaves. The stem is up to 50 cm long and bears radical leaves, red flowers in cymose heads and one-seeded white tomentose indehiscent fruits.

A sesquiterpene, jatamansone (valeranone), forms the major chemical component (up to 0.1%) of the rhizome, the drug. Other sesquiterpenoids present are spirojatamol, patchouli alcohol, norseychelanone and α-, & β-patchoulenenes. Also present are jatamol, jatamanshic acid, terpenoid coumarins oroselol and jatamansin and neolignans such as l-hydroxy pinoresinol and virolin. (Bagchi, 1991a, 1991b; Rucker et al. 1993).

Jatamansi is used as a bitter tonic, antiseptic and for treatment of epilepsy and hysteria. It is also used for insomnia and exhibit negative inotropic and chronotropic effects (Sajid et al. 1996.)

Anthemis nobilis Linn. (Asteraceae)

True Chamomile

This is a low-growing aromatic (apple-scented) herb with creeping/trailing hairy highly branched stem, perennial fibrous jointed roots, leaves which are divided to five threadlike segments, daisy like white heads (about eighteen white ray flowers around a conical center bearing yellow tubular florets) borne on erect stalks and small dry achenes. This plant is a native of Europe.

Flowers yield a volatile oil, a bitter principle (anthemic acid) and tannins. The essential oil contains α-bisacolol and chamazulene. Also present are anthesterol and apigenin.

The whole plant, especially the flower head, is used as a tonic, anodyne and antispasmodic. It is also a good emmenagogue and is an efficacious remedy for hysterical and nervous affections in women during menopause. Chamomile tea is employed as a nerve tonic and for fever and insomnia.

German chamomile is *Matricaria recutita* Rausch. containing essential oil (consists of azulene), flavones, tannins and coumarins. Aromatherapists recommend inhaling the vapourized oil of chamomile to relieve nervous disorders such as depression and anxiety associated with menopause.

Ageratum conyzoides Linn. (Asteraceae)

Sahadevi, Goatweed

Goatweed is an aromatic annual with decumbent stem, broadly ovate leaves, pale blue/white flowers (disc florets) in homogenous heads, which are borne in corymbs and glabrous achenes having awn-tipped serrate pappus scales. *A. conyzoides* is a native of tropical America, now occurs as a weed throughout.

The leaves yield an essential oil containing ageratochromene (75%), 7-methoxy-2, 2-dimethyl chromene, monoterpenes such as sabinene, β-pinene, 1,8-cineole, β-phellandrene and limonene and sesquiterpenes sesquiphellandrene, caryophyllene and caryophyllene oxide. Other constituents of the leaves are coumarin, friedelin, β-sitosterol, stigmasterol and pyrrolizidine alkaloids, lycopsamine and echinetin as well as tertiary and quaternary alkaloids. Precocenes 1 & 2 (antigonadotropic hormones), quercetin, kaempferol, 5'OMe nobiletin, sesamin, caffeic and fumaric acids and conyzorigum are the other compounds reported from leaves. The flower heads also yield a volatile oil similar to the one obtained from the leaves. The seeds contain fatty oil (25%) rich in linoleic (57%) acid (Wiedenfeld & Roder, 1991; Horiff et al. 1993).

A. conyzoides is used internally as a stimulating tonic and is useful in diarrhoea, dysentery, colic, rheumatism and fever. The juice is said to be a good remedy in *prolapsus ani.* The leaf juice is useful in boils, leprosy and skin diseases and also as an eye lotion. The essential oil is anthelmintic against tapeworm.

Artemisia cina Berg. (Asteraceae)

Kirmala, Wormseed

A. cina is a herbaceous white silky aromatic annual with lobed leaves, yellow florets in discoid heads arranged in panicles and pappus-less achenes. It is a native of South Asia.

Wormseed, the dried unexpanded flower head, contains two crystalline sesquiterpene lactones, santonin and artemisin and a volatile oil.

The wormseed or santonin is a good anthelmintic specific for roundworms and threadworms.

A number of other species of *Artemisia* are used as wormwood or sources of santonin and used likewise. They are the following:

Artemisia absinthium Linn.

Damar, Absinthe

Absinthe is an aromatic and bitter shrub with pinnatifid leaves and yellow heterogamous heads and elliptic-oblong achenes. This species is common in Kashmir. *A. absinthium* yields a volatile oil called wormwood oil or absinthe, from the leaves. This oil contains thujyl alcohol, phellandrene, pinene, *s*-guaiazulene and chamzulene. Also present in the plant 24-zeta-ethyl cholesta-7, 22-dien-β-ol, flavonoids artemisetin, homoditerpenes peroxides 1-(E)-8-isopropyl-1,5-dimethyl-nona-4,8-dienyl-4-methyl-2,3-dioxa-bicyclooct-5-ene and its diasteriomeric absinthin, anabasinthin, anabasin and thujone. Methanol extract contains α-santonin and ketopelenolid A. The powdered plant material, which forms the drug 'Afsanteen', is used in India in chronic fever, swellings, inflammation of liver and as a tonic and stimulant. The wormwood oil is applied externally to swellings and is antimicrobial in nature. Fresh wormwood is a good source of azulenes. Artemisetin exhibited marked antitumour activity against melanoma-B 16. Homoditerpenes are antimalarial substances (Rücker et al. 1992).

Artemisia maritima Linn.

Kirmala, Wormseed

This is a deciduous perennial shrub common in western Himalayas, possessing bipinnatisect leaves having many small linear segments and homogamous heads. Santonin content is maximum in unexpanded flower heads of 2-year-old plants, up to 2.2%. The essential oil (up to 0.32%) contains α-thujone (63%), sabinene, 1,8-cineole, α- and β-pinenes, camphene, bornyl acetate and citral. Aerial parts contain sesquiterpene lactones 1-oxo-6β, 7α, 11βH, 14β-methyl germacra-4 (5)-ene-12, 6-olide and 1-oxo-6β, 7α-11βH-germacra-4(5), 10(14)-dien-12, 6-olide besides vulgarin, santonin and gallicin. Kirmala is used for flavouring liquors and pharmaceutical preparations (Pathak & Khanna, 1987).

Achillea millefolium Linn. (Asteraceae)

Swetadurva, Millfoil

A. millefolium is a pubescent erect aromatic annual with stoloniferous roots, oblong, tripinnatisect leaves, pink/white flowered heads in corymbose clusters and oblong shining flattened achenes. Millfoil is found commonly in the foothills of the Himalayas.

The aerial parts of this plant are found to contain desacetylmatricarin and two sesquiterpene lactones of 3-oxyguaianolide type, 8-acetyl egelolide and 8-angeloloyl egelolide alongwith achilleine, choline, betaine, trigonelline, stachydrine, β-sitosterol, and a volatile oil. The oil, maximum in heads, contains three antitumour sesquiterpenoids, achimillic acids A,B&C, isoapressin, 10-isovaleryl desacetylisoapressin, 8-tigloyldesacetylisoapressin, 8-tigloyldesacetylezomontanin, α-peroxyachifolid, β-peroxyachifolid, guaianolide, azulene lactones such as millefin, *de-Ac*-matricarine, chamazulene, α-pinene and other minor constituents. Leaves yield luteolin, apigenin, folic acid, rutin and ascorbic acid (Rükker et al. 1991).

Milfoil is astringent, tonic, diaphoretic and anti-spasmodic and is useful in treating cold, colic, hysteria, epilepsy and rheumatism. Infusion of the plant is useful in preventing premature ageing of the skin. Achilleine is reported to be haemostatic and reduce blood clot time in rabbits.

Santalum album Linn. (Santalaceae)

Safed Chandan, Sandalwood

Sandalwood is the heartwood of a small tree native to peninsular India. This is a root parasite with glabrous elliptic ovate leaves, small purple/violet flowers in terminal or axillary paniculate cymes and purple black globose drupes.

The oil, amounting to 4-5% of the wood, is extracted by water distillation, which is further purified by a subsequent distillation. The roots are richer in oils, yielding up to 10% dry weight. The refined oil is a pale yellow clear liquid consisting principally of a sesquiterpene alcohol, santalol (90%). Santalol exists both in α- and β-forms, the former being predominant. Also present in the oil are santene (C_9H_{14}), α- and β-santalenes and santalone.

Sandalwood oil is one of the very important raw materials in perfumery, soaps, face creams, powders, etc. It has a very high fixative value due to its high boiling point. The wood is a sedative, diuretic, and tonic and useful in bronchitis, biliousness, thirst, cystitis, gonorrhoea, etc. The wood paste is applied externally for ulcers, headache, inflammation and pruritus. This wood is ideal for carving.

Cichorium intybus Linn. (Asteraceae)

Kasani, Chicory

This is an erect perennial with a fleshy cylindrical long (up to 75cm) tap root, broadly oblong/oblanceolate lower leaves crowded at the base forming a rosette, cordate upper leaves, blue flowers in homogenous heads and 5-angled achenes. A native of Europe, chicory is now cultivated everywhere.

The fresh roots contain inulin, gum and sesquiterpene lactones such as lactucin, 8-deoxylactucin, lactucopicrin, crepidiaside, sonchusides, cichoriolides and cichoriosides. Also present are ribosylzeatin, chlorogenic, neochlorogenic, isochlorogenic, caffeic and chicoric acids. Inulin, on storage, gets converted to fructose.

Chicory is used mixed with coffee. The root is a carminative, diuretic and tonic and is effective in jaundice, liver enlargement, gout and rheumatism. The whole plant is used in fevers, diarrhoea and as emmenagogue and alexiferic.

Applied externally as a poultice, roots and leaves cure inflammatory swellings. The whole plant is a good antiseptic. Seeds yield 'Ben oil' which is a valuable lubricant.

Pogostemon cablin Benth. (Lamiaceae)

Syn. *P.patchouli* var. *sauvis* Hook.f.

Patchouli

Patchouli is an erect much-branched aromatic herb (upto 1.2m high) with ovate crenate-serrate tomentose leaves, white flowers with purple streaks in whorls on terminal and/or axillary spikes. This plant is a native of Philppines.

Leaves yield the volatile patchouli oil (upto 5.8% dry wt basis) and a resin. The oil consists of sesquiterpenes, patchouli alcohol (major compound), β-patchoulene, α-guaiene, α-bulnesene and benzaldehyde.

The leaves and tops of Patchouli are antirheumatic and said to be good for menstrual problems.

Saussurea lappa C.B. Clarke (Asteraceae)

Kushta, Costus

This is a woody erect perennial reaching a height of 2m, with stout robust pungent roots (up to 60cm long), fibrous thick stem, large (up to 1m) radical leaves having long lobately winged stalks, large flower heads (3-5cm in diam.) containing dark blue-purple or black flowers and small curved compressed achenes. Costus is endemic to Kashmir valley at altitudes of 2500-3000m.

The roots, which form the drug, yield a resin (6%), volatile oil (1.5%), inulin (18%) and alkaloids. The oil consists of costunolide (a sesquitepene lactone-principal component), dehydrocostuslactone, dihydrocostuslactone, 12-methoxy dihydrocostunolide, costol and β-ionone. β-Sitosterol, stigmasterol and betulin are the components of resin. The alkaloids reported are saussurine and kushtin.

Kushta is a well-known incense, insecticide and used in asthma and chronic bronchitis.

Anamirta cocculus W. & A. (Menispermaceae)

Kakamari, Fish Berries

A. cocculus, a native shrub of India, yields the 'fish berries'. The plant is a climber with glabrous corky shoots, ovate leaves, unisexual flowers in panicles and a drupe containing globose seeds.

The seeds contain approximately 1.5% of a bitter principle picrotoxin (cocculin) consisting of equimolecular proportions of highly oxygenated sesquiterpenes picrotoxinin ($C_{15}H_{16}O_6$), picrotin ($C_{15}H_{16}O_6$) and about 50% fat.

Kakamari seeds are highly toxic to fish. Picrotoxin is a potent long-acting respiratory stimulant and used to treat overdoses of CNS depressants such as barbiturates and narcotics.

Elephantopus scaber Linn. (Asteraceae)

Gojiva, Elephantopus

Elephantopus is a rigid perennial herb with short rootstock with numerous stout fibrous roots and a dichotomously branched short stem covered by appressed white hairs. Leaves are mostly radical, oblanceolate-spathulate forming a spreading rosette on the ground; cauline leaves smaller and sessile. The flowers are violet in homogamous heads of 2-5 flowers, the heads collected into a closely packed terminal inflorescence, the achenes are truncate, 10-ribbed. This is a common weed forming undergrowth in the Western Ghats.

The plant contains hydroxylated germacrolides, molephantin, molephantinin, phantomolin and its *cis*-epoxide, elephantin, elephantopin, deoxyelephantopin, *iso*-deoxyelephantopin, 11,13, dihydrodeoxyelephantopin, along with α-curcumene, β-amyrin. lupeol, epifriedelinol and stigmasterol. Also present are 4,5- dicaffeoyl quinic and 3,5-dicaffeoyl quinic acids. Roots contain deacylcyanopicrin, glucozaluzanin-C and stigmasterol (Hisham et al. 1992).

The roots, which form the drug, are useful in fever, cardiac problems and liver trouble. This plant is also used in insomnia, diabetes, and urethral discharges, rheumatism and for filariasis. Molephantin and molephantinin possess cytotoxic and antitumour properties, the latter also shows antileukemic properties. Phantomolin and its *cis*-epoxide exhibit potent inhibitory action on Ehlich ascitis carcinoma and Walker 256 carcinosarcoma cells.

Inula racemosa Hook.f. (Asteraceae)

Puskarmula, Inula

I. racemosa is a tall stout herb with a grooved stem, many radical broad elliptic-lanceolate leaves (tomentose beneath) heterogamous yellow/white heads and achenes having red pappus. Inula is common in Jammu and Kashmir and Himachal Pradesh.

The roots yield an essential oil, a germacranolide, i.e. inunolide, alantolactones such as alantolactone and isoalantolactone and β-sitosterol.

Roots are used as a general tonic, aphrodisiac, diuretic and febrifuge. The drug is also used for cough, anaemia, cardiac disorders, etc. Root extract is found to be anti-inflammatory, antipyretic, anthelmintic, and hypolipidic.

Inula helenium Linn.

The roots of this plant is used for the same purposes as *I. racemosa* in Europe. They yield alantolactones, whereas the aerial parts contain scopoletin and umbelliferone.

Cyperus rotundus Linn. (Cyperaceae)

Musta/Nagar Moth

This is a perennial herb with stoloniferous rhizome bearing hard, ovoid, tunicate, black fragrant tubers (up to 3 cm in dia.) triquetrous stem, flat 1-nerved narrowly linear finely acuminate leaves, spikes in umbels or contracted into a head, and trigonous black nuts. Musta is a very troublesome weed in most tropical countries.

The tubers yield an essential oil and steroids like oleanolic acid, its glycoside and β-sitosterol. The volatile oil is rich in sesquiterpenes such as patchoulenone, cypernone, α- and β-retunol, cyperotundone and nor-sesquiterpenes, kobusone and isokobusone.

Musta is used for indigestion, diarrhoea, sprue and other stomach problems. It is diuretic, carminative, anthelmintic, galactogogue, emmenagogue and nervine tonic. It is also useful for epilepsy, increasing memory power and a tonic.

Glucosinolates and Sulphides

Lepidium sativum Linn. (Brassicaceae)

Halim, Garden Cress

This is a small herbaceous glabrous annual with variable leaves (sessile linear/pinnatifid cauline leaves, long-petioled bipinnate radical leaves) small white flowers in racemes and small orbicular, ovate notched pods containing two winged seeds. A native of Ethiopia, garden cress is cultivated as a salad plant everywhere.

The leaves contain protein (5.8%), carbohydrates (8.7%) and a glucosinolate glucotropaeolin which on steam distillation produces a pungent volatile oil (cress oil, 0.115%) consisting of benzyl isothiocyanate and benzyl cyanide. Seeds contain protein(24%), fat (16%), glucotropaeolin, sinapin (choline ester of sinapic acid), sinapic acid, mucilage and an alkaloid. The fixed oil is rich in oleic (62%), and linolenic (28%) acids and contains β-sitosterol and α-tocopherol. Mucilage is an acidic polysaccharide consisting of galacturonic acid, L-arabinose, galactose, rhamnose and glucose.

Garden cress is consumed raw in salads. It is used for treating asthma, cough and bleeding piles. Leaves are also used as a diuretic, for scorbutic diseases and liver complaints. Seeds are rubefacient, galactogogue, emmenagogue, laxative, tonic, aphrodisiac and diuretic. Roots are used in syphilis and tenesmus.

Tropaeolum majus Linn. (Tropaeolaceae)

Garden Nasturtium

This is a small spreading succulent herb having peltate, orbicular zig-zagged long-petioled leaves, large zygomorphic spurred red/scarlet (or variously colored) flowers and 3-seeded schizocarp. A native of South America, this plant is now grown everywhere for the colourful flowers and beautiful foliage.

The herb yields a water-soluble essential oil (on steam distillation) consisting mostly of benzyl isothiocyanate. The glucosinolate present in the plant is glucotropaeolin which is acted upon by myrosinase to give benzyl isothiocyanate. Also present in the leaves are vitamin C and glycosides of quercetin and phenolic acids. Seeds yield protein (27%), fixed oil (10%-containing 72% erucic acid and 21% eicosenoic acid) and a volatile oil consisting entirely of benzyl isothiocyanate. Flowers contain carotenoids, kaempferol and pelargonidin.

The herb is employed in cystitis and for inflammation of kidneys. The extracts are strongly antibiotic and used for urinary and respiratory diseases and eye diseases. Leaves and petioles are used in salads and the pungent fruits are edible.

Armoracia rusticana Gaertn. (Brassicaceae)

Syn. *A. lapathifolia* Gilib.

Horse radish

This tall hardy glabrous plant possesses a conical woody/fleshy root, large stems, large oblong lower leaves, small lanceolate upper leaves, white flowers in racemes and globose pods containing many small seeds. It is a native of S. E. Europe, now cultivated in hilly areas of India.

The thick roots, used as a condiment, possess a bitter taste and a characteristic pungent odour due to allyl isothiocyanate, occurring free or as the glucoside, sinigrin. Horse radish contains a volatile oil, protein (2%), sugars (3%) and vitamin C. The oil, when isolated, consists of allyl and/or 2-phenethyl isothiocyanate as major constituents along with methyl, isopropyl, 2-butyl or 4-pentenyl isothiocyanate. The minor components are considered to be produced as artifacts during extraction.

Capsella bursa - pastoris Medic. (Brassicaceae)

Mumiri, Shepherd's Purse

This is a pubescent erect annual with radical pinnately divided leaves, small cauline leaves having hastate base, white small flowers in racemes and glabrous obcordate, laterally compressed capsules enclosing numerous ellipsoid reddish brown seeds. Occurs as a weed throughout the temperate regions.

The plant is found to contain flavonoids, glucosinolates, volatile oil and saponins besides choline. Choline amounts to 0.2%. The flavonoids reported are hesperidin, rutin, robinetin, derivatives and glycosides of luteolin, quercetin, kaempferol, fisetin, gossypetin and related compounds. The sulphur containing compounds present are sinigrin and α-methylsulfinylnonyl and 10-methylsulfinyl decylglucosinolates, thiocyanic acid and s-methyl-L-cysteine sulphoxide. The volatile oil is rich in camphor. Other compounds isolated are coumarins, ascorbic acid, histamine, two crystalline alkaloids, tyramine, inostiol, a proteolytic enzyme, β-sitosterol and stigmasterol.

Seeds yield oil (26%), protein (32%), mucilage, sulphur containing volatile oil and quercetin. The oil consists of linolenic (35%), linoleic (18%), gadoleic (13%), oleic (11%) and palmitic (9%) acids.

The herb possesses bitter, pungent, astringent, antiscorbutic, febrifugal and emmenagogue properties and is substituted for *Hydrastis*. This plant is a remedy for atrophy of limbs by rubbing the affected parts with an alcoholic solution of *Capsella* and *Alchemilla vulgaris*. The juice of the plant checks menorrhagia and other haemorrhage of renal and urino-genital tracts. The leaves are used as salads and its seeds are used for bread making. The seeds possess rubefacient and vesicant properties.

Moringa oleifera Lam. (Moringaceae)

Syn. *M. pterygosperma* Gaertn

Sigru, Drumstick

Drumstick tree is a soft-wooded medium-sized tree with pungent roots, large 3-pinnate leaves having articulated rachis, small elliptical (terminal leaflet obovate) leaflets, white flowers (with reflexed sepals and spathulate petals) in axillary or terminal panicles and a long-pendulous cylindrical 9-ribbed capsule constricted between seeds and enclosing trigonous seeds broadly winged at the angles. This is found wild in western Himalayas and grown all over India and tropics for its edible leaves, flowers and fruits.

The root bark contains three alkaloids moringine (benzylamine), moringinine and spirochin and glucosinolate pterygospermin yielding benzyl isothocyanate. Leaves contain kaempferol, 3′-OMe quercetin, acacetin, vitexin, vanillic acid, syringic acid and melilotic acid. Fruits contain proanthocyanidins (Daniel, 1989).

All parts of the plant are useful. Roots and root bark are used in fever, asthma, rheumatism, inflammation, epilepsy and hysteria. Leaves are galactogogue, useful in eye diseases, fainting fits and spasmodic problems of bowels and hysteria. Roots, leaves, fruits and seeds are attributed with a number of medicinal properties such as aphrodisiac, cardiotonic, anthelmintic and good for eye diseases, enlargement of spleen and joints. Applied externally as a poultice, the roots and leaves cure inflammatory swellings. The whole plant is a good antiseptic. Seeds yield 'Ben oil' which is a valuable lubricant.

Allium sativum Linn. (Liliaceae)

Lahsuna, Garlic

A. sativum is a hardy perennial with an underground bulb consisting of a disc-shaped stem, swollen axillary buds (cloves) encircled by 2-3 layers of tunica and long flat leaves. Flowers are small, white, borne on scapes. A native of Central Asia, garlic is now cultivated everywhere.

Garlic bulbs contain carbohydrates (30%), protein (6%) and an essential oil (0.1%) besides minerals and vitamins. Also present are peptides such as γ-glutamyl phenylalanine, γ-glutamyl-S-methyl cysteine, γ-glutamyl-S-β-carboxy-β-methyl ethyl cysteinyl glycine, S-allyl mercapto-L-cysteine, γ-glutamyl-S-allyl mercapto-L-cysteine and γ-glutamyl-S-propyl cysteine. In addition garlic contains sulphur containing amino acids such as S-propyl-L-cysteine, S-propenyl-S-cysteine, S-methyl cysteine, S-allyl cysteine, S-allyl cysteine sulphoxide (alliin, which is acted upon by an accompanying enzyme allinase produces allicin, i.e. thio-2-propene-l-sulphinic acid-S-allyl ester) and S-butyl L-cysteine sulphoxide. Other compounds reported from garlic are a thioglucoside, scordine, scordinines A, A2 and B and cyanidin. The volatile oil of garlic consists of allyl alcohol, allyl propyl disulphide, methylallyl disulphide, diallyl disulphide, dimethyl trisulphide, methyl allyl trisulphide and diallyl trisulphide. Garlic skin is rich in pectin.

Garlic is a stimulant, diaphoretic, expectorant, diuretic, and tonic. It exhibits anthelmintic, emmenagogue and antidysenteric properties.. It is also used as anti-tubercular, anticholeric, anti-fertility, hypocholesteremic and hypoglycemic drug. Almost all the compounds from garlic exhibit anti-microbial activity. Allicin exhibits anticancerous properties and is used in treatment of rheumatoid arthritis. The volatile oil is used externally in paralytic pain. It is found to increase fibrinolytic activity in patients with coronary artery disease. Ajoene, a derivative of allicin, is antileukaemic in nature.

Other species of *Allium* used in medicine are the following.

Allium cepa Linn.

Piyaz, Onion

A. cepa is used as a vegetable and contains an essential oil (rich in allylpropyl disulphide), soluble carbohydrates, vitamins, γ-glutamylpeptides, phenolic acids, flavonols, anthocyanins, and oleanolic acid. The anthocyanins are 4-substituted, carboxypyranocyanidins such as 5-carboxy-2- (3,4-dihydroxy phenyl)-3,8-dihydro pyrano-(4,3,2-de)-1-benzopyrillium (Fossen and Andersen 2003). The juice of onion is used for treatment of ophthalmia, earache and jaundice. Onion is also said to possess aphrodisiac, antimalarial and anti-rheumatic properties. It is also a hypoglycemic agent.

Allium porrum Linn.

Vilayati piaz, Leek

Leek is used as a substitute of garlic and contains more (up to 17%) carbohydrates.

Allium schoenoprasum Linn.

Chives

The young leaves and bulbs of this plant, which are eaten as salad, are found to exert hypotensive and cardiac depressant properties. The leaves yield a volatile oil (consisting of sulphides) and a number of peptides.

Ferula asafoetida Bios (Apiaceae)

Hing, Asafoetida

F. asafoetida and related plants, natives of Iran, India and Afghanistan, yield the oleo-gum-resin asafoetida. *F. asafoetida* is a stout perennial herb 2-3 m in height with 2-4 pinnate pubescent leaves having cauline sheaths, yellow flowers in large terminal compound umbels and dorsally compressed cremocarps. Cutting off successive horizontal slices collects the gum resin, occurring in the cortex of thick fleshy rootstock, after removal of the crown. The fresh resin oozing out and collected at the cut surface is yellowish white and translucent, but becomes a reddish brown opaque solid on standing.

Asafoetida possesses a powerful alliaceous odour and a bitter acrid taste. It contains a volatile oil (4-20%), resin (40-60%), gums (20%) and some impurities. The oil consists chiefly of isobutyl propanyl disulphide and related compounds, which are responsible for the characteristic smell of the gum resin. The resin is composed of coumarins such as umbelliferone, 5-hydroxyumbelliprenin, 8-hydroxyumbelliprenin, 9-hydroxyumbelliprenin, 8-acetoxy, 5-hydroxy umbelliprenin, asafoetidin, ferocolicin and asacoumarins A&B; phenols such as asaresinol ferulate and free ferulic acid and a group of sesquiterpenes, farnesiferol A, B, & C characteristically containing coumarin groups (Appendino et al. 1994).

Asafoetida is extensively used in the east for flavouring foodstuffs and in the west as an ingredient of perfumes in dilute concentration. Medicinally, it is used as a carminative, expectorant, antispasmodic, laxative and against asthma, whooping cough, chronic bronchitis, hysteria, epilepsy and cholera.

3.5 Diterpenes

Diterpenes are C20 compounds formed by the condensation of four isoprene residues. The known 1000 or more diterpenes fit into 20 major common skeletons and four less common skeletons. Similar to the other terpenes, hydrocarbons, alcohols, aldehydes, ketones, and acids are all known in this group. Phytol, which constitutes the tail of the chlorophyll molecule, is the only acyclic diterpene known and most of the diterpenoids are regarded to be derived from it by ring closures, oxidation and substitutions. Artemisone, wormwood diterpene and camphorene represent a few monocyclic derivatives known. The rest of diterpenes have larger ring structures. The important skeletons found in this category are: (1) labdane (bicyclic), (2) abietane, (3) pimarane, (4) cassane (tricyclic), (5) kaurene, (6) gibbane, (7) bayerene (tetracyclic), and trachylobane (pentacyclic). Some of the physiologically active diterpenes are: (1) gibberellins (growth regulators), (2) ryanodine (insecticide) and (3) phorbol (purgative).

Diterpenes exhibit some very interesting properties. Stevioside, a diterpenoid glycoside isolated from leaves of *Stevia rebaudiana* Bert. (Asteraceae), is at least 300 times sweeter than sugar and used as a sweetener for diabetic patients and dieters. Bitterness is a phenomenon of certain diterpenoids such as marrubin and columbin. A number of other diterpenes are isolated from plants showing a variety of properties. One of the most acclaimed antileukemic principle, taxol, isolated from *Taxus brevifolia* Nutt. is a diterpene. Other diterpenes exhibiting antitumour activity are the following:

1. Taxodone and taxodione from *Taxodium distichum* (Taxodiaceae)
2. Nagilactones B - E from *Podocarpus* sp. (Taxaceae)
3. Podolide from *P. gracilior*
4. Triptolide and tripdiolide from *Tripteryguim wilfordii* (Celastraceae)
5. Gnidin, gniditrin, gnidicin from *Gnidia lamprantha* (Thymeliaceae)
6. Mezerein from *Daphne mezereum* (Thymeliaceae)
7. Jatrophone from *Jatropha gossypifolia* (Euphorbiaceae)
8. Jatrophatrione from *J. macrorhiza*
9. Ingenol-3, 20-dibenzoate from *Euphorbia esula* (Euphorbiaceae)
10. Phorbol, 12-tiglate, 13-decanoate from *Croton tiglium* (Euphorbiaceae) (Sticher, 1977)

Ginkgo biloba Linn. (Ginkgoaceae)

Ginkgo, Maidenhair Tree

G.biloba is a tall dioecious tree with fan-shaped long-petioled obconical bilobed leaves having dichotomous venation, borne in whorls in short shoots (spurs). Staminate trees have stamens in pairs borne in catkin-like bractless cones, present in leaf axils. Ovulate trees bear ovules in pedunculate pairs (one of each pair often aborting) on short shoots. The so-called 'fruit' is actually a seed, plum like and drupaceous, having a fleshy outer layer and a horny inner layer ('seed'). A native of China, *G. biloba* is occasionally cultivated in gardens in many places.

Leaves contain terpene lactones (6%) and flavonols (24%) as the major constituents. The terpene lactones include diterpenoids ginkgolides A, B and C and a sesquiterpene, bilobalide A. The flavonols present are kaempferol, quercetin and isorhamnetin as mono-, bi- and triglycosides and/or as esters with coumaric acids. Biflavones seen are 1,5-di-methoxy bilobetin, amentoflavone, lobstein, ginkgetin, isoginkgetin and sciadopitysin. Procyanidin, catechins, shikimic acid, sequoyitol and pinitol are the other compounds reported. Seeds contain bilobalide, ginnol, bilobals, cardenols and phenolic lipids, 4-hydroxy anacardic acids like 6-(pentadec-9-enyl) resorcylic acid and 6-tridecyl resorcylic acids (Joly et al. 1999).

In traditional Chinese medicine, seeds are considered more important whereas in western countries, leaves are held in high esteem. Ginkgo is used generally for diseases appearing due to advanced age. Leaves are effective in improving memory, removing depression and used in cardiovascular disorders, Alzheimer's and Parkinson's diseases, cerebral ischemia and liver fibrosis. It is vasodilatory, aphrodisiac and modulate cerebral energy metabolism due to antioxidant properties of flavonoids. The flavonoids and ginkgolide B act as a platelet-activating factor (PAF) antagonists, strengthen blood capillaries in brain and ensure that all parts of the brain get proper oxygen through blood. This and the antioxidant, free-radical scavenger properties of flavonoids maintain proper cerebral functioning. Since PAF is linked with many cardiovascular, renal, respiratory and CNS disorders, all these diseases are prevented by this drug. It is also useful in premenstrual syndrome, in acute allergic inflammations and increases the blood flow to the retina and reduces retinal deterioration and hearing loss.

The seeds are used as an expectorant, for bladder ailments, spermatorrhoea, menorrhoea, uterine fluxes and cardiovascular ailments.

Taxus brevifolia Nutt. (Taxaceae)

Syn. *T. baccata* subsp. *brevifolia* Pilger

Pacific Yew

This is a small dioecious tree, with scaly, purplish bark, drooping branches, small, linear, acute, mucronate leaves having a decurrent leaf base, arranged in two rows, staminate flowers solitary or in small axillary cone like clusters of 3-14 stamens (anthers 3-9 celled) borne on peltate or apically thickened microsporophylls and female cone of a solitary terminal ovule borne on a fleshy fertile sporophyll. The seed is dark blue, ovoid, 2-4 angled, surrounded by a red fleshy cup-shaped cupule. Pacific yew is a native of America.

The bark yields a group of diterpenes (taxanes), alkaloids (taxine and taxagitine), lignans (isotaxiresinol), flavonoids (quercetin, rhamnetin and sciadopitysin) and glycosides. Taxanes isolated include taxol (an amide), baccatin, brevifoliol, cephalomanine and many derivatives of taxol.

Taxol (also known as paclitaxel) is one of the most potent anticancer compounds today. It disrupts microtubule formation by binding to the tubulin to form abnormal mitotic spindles and, thus, stops cell division. It is also used in chronic fatigue syndrome, herpes, lupus and candida. Nearly 1000 kg of bark are needed to produce 1 kg of taxol.

Taxus baccata Wall., the Himalayan yew, provides diterpenoids like 10-deacetylbaccatin, from needles and twigs, which can be semi-synthetically converted to taxol. Taxol is obtained from *Taxus chinensis* also, which is a richer source (upto 0.02%). It is also found that *Taxomyces andreanae*; an endophytic fungus occurring in *Taxus brevifolia* also produces taxol. Recently, it has been found that the hazelnut tree (*Corylus avellana*) also contains taxol.

Swertia chirata Buch. -Ham. (Gentianaceae)

Kairata

This is a small stout herb branching towards the tip, with a green stem, opposite sessile elliptic 7-nerved leaves, greenish-yellow flowers (petal lobes with two glands on either side) in axillary/terminal corymbose cymes and sessile oblong capsule enclosing numerous minute ovoid bitter seeds. Kairata is found in temperate Himalayas.

The entire plant, used as a drug, contains bitter principles amarogentin (0.04%) and amaroswerin (0.03%) as major components. Other compounds present are xanthones such as chiratol, methyl bellidifolin, decussatin, 7-OMe swertianin, mangiferin, swertianin, swertinin and chiratanin; triterpenoids like masilinic acids (and its glycoside swericinctoside), chiratenol, gammacer-16-en-3β-ol, 21α-H-hop-22(29)-en-3β-ol, swertenol, episwertenol, pichierenol and kairatenol; secoiridoid glycosides, viz., swertiamarin and gentiopicroside and alkaloids, such as gentianine, gentiocrucine and enicoflavine (Sharma, 1982; Mandal and Chatterjee, 1987; Astham et al. 1991; Chakravarty et al. 1992).

Kairata is a bitter tonic and febrifuge. It is also used as antihepatotoxic, antiinflammatory, antiulcerogenic, anticholinergic, CNS depressant, antimalarial and hypoglycemic.

Andrographis paniculata Wall. (Acanthaceae)

Kalmegh, Creat

A. paniculata is a small erect annual with winged quadrangular stem, glabrous lanceolate leaves, small white purple flowers borne on horizontal axillary racemes and linear-oblong capsules containing numerous subquadrate brown seeds. Kalmegh is a native of tropical Asia.

The whole plant is intensely bitter due to the presence of diterpenoid lactones such as andrographolide (0.6%), 1, 4-deoxy-11-oxoandrographolide, 14-deoxy-11,12-didehydroandrographolide, 14-deoxy andrographolide and a non-bitter neoandrographolide. Also present are flavones such as 5-OH,7,2′6′-triOMe flavone, 5-OH,2′,7,8 triOMe flavone, 2,5-diOH-7,8-diOMe flavone, mono-O-methyl wightin, 5-OH-7,8-diOMe flavone and 5-OH-3,7,8,2′-tetraOMe flavone; a C_{23} terpene, 14-deoxy-15-isopropylidine-11,12-didehydroandrographolide and an iridoid glucoside-procumboside. The leaves contain up to 1% andrographolide. The roots, in addition to andrographolide, contain flavones such as 7,4′-diOMe apigenin, echioidin, andrographin and panicolin, as also α-sitosterol (Abeyasekara et al. 1990; Reddy et al. 2003).

This herb is extensively used as a bitter tonic and febrifuge. It is also used in dysentery, cholera, diabetes, influenza, piles and gonorrhoea. It is also considered a tonic, blood purifier and a cure for liver dysfunction.

Aleurites fordii Hemsl. (Euphorbiaceae)

Tung Oil tree

A small, much-branched milky tree, *A. fordii* possesses simple trifid leaves, unisexual flowers in clusters in which one or more pistillate flowers are surrounded by a number of staminate flowers, spheroid/top-shaped fruits containing 3-5 one-seeded segments. It is a native of China but currently cultivated widely in India.

The leaves yield three ellagitannins, aleurinins A,B&C along with corilagin, geraniin and chebulagic acid. This wood is found to contain a coumarino-lignoid and 5,6,7-trimethoxy coumarin. Seeds yield 30-35% oil containing α-elaeostearic (75-85%) and linoleic acids (10%) as the principal fatty acids, about 10% proteins, starch and 2 diterpene esters 13-O-acetyl-16-hydroxy phorbol and 12-O-palmityl-13-O-acetyl-16-hydroxy phorbol and a non-protein amino acid, L-3-carboxyl-1, 2, 3, 4-tetrahydro-β-carboline. The bark and fruit husk contain 12 and 4.5% tannins, respectively (Fozdar et al. 1989).

The seed oil is used for treating boils, ulcers, swellings and burns. Phorbol derivatives are anticancerous.

Baliospermum solanifolium Suresh (Euphorbiaceae)

Syn. *B. montanum* Muell. Arg
B. axillare Blume

Danti, Baliospermum

A stout undershrub of about 2m, with many shoots arising from the base of stem carrying large (lower) and small (upper) palmately lobed leaves, small green monoecious flowers in axillary racemes (male flowers more common, towards the top; female flowers few at the base), and 3-lobed obovoid capsules containing oblong, mottled, shiny seeds. *B. solanifolium* is common in all parts of India.

Roots contain phorbol derivatives such as montanin, baliospermin, 12-deoxyphorbol palmitate, 12-deoxy-16-hydroxy phorbol palmitate and 12-deoxy-5β-hydroxy phorbol myristate. The seeds yield an oil with a hydroxy fatty acid, i.e. 11,13-dihydroxy-tetracos-*trans*-9-enoic acid.

The roots are considered pungent, anthelmintic, diuretic and useful for piles, enlarged spleen and skin diseases. It is also cathartic and used in dropsy and jaundice. Ethanolic extract of roots are active against P-388 lymphocytic leukaemia. Whole plant is used for treating cancer and abdominal tumours. The leaves also exhibit purgative and expectorant properties and useful in dropsy, while the seeds are drastic purgatives. Seed oil is a powerful hydragogue and used externally in rheumatism. Seeds are substituted for those of *Croton tiglium* (Jamalgota).

Jatropha gossypifolia Linn. (Euphorbiaceae)

Dravanti, Bellyache bush

This is a tall shrub with branches and petioles clothed with numerous fascicled and branched gland-tipped bristles. The leaves are simple, brown when young, turning green on maturing, lamina deeply 3-5 lobed and margin glandular hairy. Flowers are unisexual, and red in terminal cymes, capsule is 3-lobed. A native of Brazil but now dravanti has become naturalized in India and is seen in all parts of India.

The roots contain macrocyclic diterpenes, jatrophone and jatropholones A & B, and the bark yields β-sitosterol. The leaf contains flavones and lignan.

Seeds are used as drastic purgatives, aphrodisiac, anthelmintic and are useful in piles, enlarged spleen and skin diseases. Seed oil is used in rheumatism and is a remedy for itch, herpes and eczema. Leaves are galactogogue, insecticidal and used for wounds and ulcers. Roots and leaves are also used in dysentery, anaemia, biliousness, fistula and ulcer.

Jatropha glandulifera Roxb.

Dravanti

A plant similar to *J.gossypifolia* with smaller leaves having serrate margins, this is a native of Sri Lanka, now common as a wild plant in many localities. It is used in place of Dravanti in many places.

The latex and leaves are used in warts and tumours. The plant yields pigments such as 3,3-dimethyl acrylylshikonin and isohexenylnaphthazarin

Boswellia serrata Roxb. (Burseraceae)

Shallaki/Salai, Frankincense

B. serrata is a medium-sized deciduous aromatic tree with smooth greenish resinous bark exfoliating in thin papery scales, moderately large imparipinnate leaves having ovate-lanceolate serrate leaflets, small white flowers in axillary racemes, trigonous drupes splitting along 3 valves and compressed winged seeds. A native of India, shallaki is very common in central parts of India.

On injury from the bark, an oleo-gum-resin oozes out, which on exposure to air hardens to brownish yellow tears or crusts known as frankincense (salai guggul). The resin is also extracted by tapping the bark. It contains 8-9% volatile oil, 40-45% resin and 30-35% gum. The oleo-gum-resin is treated with pure ethanol with vigorous shaking when the gum gets precipitated which is then filtered off. The ethanol solubles are concentrated and steam distilled to remove the volatile oil. The resin remaining in the flask is taken out by dissolving in ether or alcohol.

The volatile oil consists of α-thujene (50%), and *p*-cymene (14%) as major components and β-pinene, *d*-limonene, linalool, terpeneol, terpenyl acetate, methyl chavicol, cadinene, geraniol and elemol as minor components. Also present in the oil are three terpenoic acids, viz., α-campholenic acid, α-campholytic acid and 2,2,4-trimethylcyclopent-3-en-l-yl- acetic acid. The resin consists of a diterpene alcohol, serratol and eight triterpene acids such as derivatives of tirucallenoic acids (four in number, viz., 3-α-acetoxy tirucall-8, 24, dien-21-oic acid, 3-ketotirucall-8, 24-dien-21-oic acid, 3-α-hydroxy tirucall-8, 24-dien-21-oic acid and 3-b-hydroxytirucall-8, 24-dien-21-oic acids) and four boswellic acids (β-bosewellic acid, acetyl β-boswellic acid, acetyl-11-keto-β-boswellic acid and 11-keto-β-boswellic acid) besides α- and β-amyrins. The gum is an acidic polysaccharide consisting of uronic acid (30%), galactose (46%), D-arabinose, xylose and mannose (Pardhy and Bhattacharya, 1978).

The bark is found to contain tannins (9%) and β-sitosterol. Seeds contain protein (9%) and carbohydrates (30%). The leaves contain methoxy quercetins, myricetin, proanthocyanidins, gallotannins and phenolic acids (syringic, gentisic, and gallic acids).

Frankincense is very effective in osteoarthritis, juvenile rheumatoid arthritis and spondylitis and is used as astringent, stimulant, expectorant, diuretic, diaphoretic, emmenagogue, and ecbolic and also exhibits antispetic properties. It is also useful in ulcers, tumours, rheumatism, inflammations, piles and skin diseases.

Caesalpinia bonduc Roxb. (Caesalpiniaceae)

Kuberakshi, Bonduc nut/Fever nut

A heavily armed woody climber, *C. bonduc* possesses branches and rachis of leaves armed with stout recurved prickles, bipinnate leaves with 6-11 pairs of ovate leaflets and foliaceous stipules, yellow fragrant flowers in the racemes and swollen flattened spiny beaked pods containing 1-2 large orbicular shiny polished seeds. A native of India and Persia, it is found wild in many parts of India.

The seeds contain proteins (25%), oil (20-24% rich in linoleic and oleic acids) and α-, β-, γ-, δ- and ε-caesalpins, caesalpin F and a homoisoflavone, bonducellin. The leaves contain protein (25%), a bitter principle ($C_{24}H_{32}O_8$), wax and pinitol.

Bonduc nuts are known for their antiperiodic efficacy. They are used as diuretic, antipyretics, and tonic, in malaria, diarrhoea and for treating hydrocele. The seed oil is antirheumatic in nature. Leaves are used in skin diseases and rheumatism. Roots are used to cure leucorrhoea, in intermittent fevers and diabetes.

Caesalpinia crista Linn.

Syn. *C. nuga* Ait

C. crista is considered same as *C. bonduc* by many, but differs from the latter in having simple stipules and smooth pods (Sivarajan and Balachandran, 1994). Since the identity of the plants studied is controversial, the compounds reported from *C. bonduc* would have been from the material of *C. crista* and vice versa. Therefore, the chemistry and properties of both the plants, for the present, are considered to be the same.

Calophyllum inophyllum Linn. (Clusiaceae)

Nagchampa, Alexandrian Laurel

C. inophyllum is a moderate-sized evergreen tree with milky latex, black bark, large, broadly elliptic oblong leaves, fragrant polygamous white flowers in axillary racemes and light yellow drupes containing a single large globular seed. This plant is a native of S.E. Asia, common in the coastal regions of India.

The seeds contain large amounts of oil (60%) of a disagreeable odour, protein (6%) and carbohydrates (4%). The oil consists of oleic (60%), palmitic (20%) and stearic acids (13%). Other constituents of oil are calophyllolide, calcustralin, inophyllolide, apetalolide and calophyllic, inophyllic, inophenic, calophynic and pseudobasinic acids. Cinnamic acid, an essential oil and leucocyanidin are the other constituents of the seed. The root bark contains xanthones, caloxanthones A,B & C, 4-OH xanthone, 1-OH, 2OMe xanthone, 1,2-diOMe xanthone, macluraxanthone, and 1,5 diOH xanthone, besides 6-deoxyjacareubin and (-)epicatechin. Other compounds extracted are inophyllums A,B,C,E,P,G_1 and G_2. The heartwood contains xanthones like calophyllin B and 1, 5, 6-trihydroxy xanthone, jacareubin, 6-deoxyjacareubin, 1,3,5,6-tretrahydroxy xanthone, 2-(3-methyl but-2-enyl)-1,2,5,6-tetrahydroxy xanthone, 2-(3-methyl but-2-enyl)-1,3,5-trihydroxy xanthone and 2-(3-hydroxy-3-methyl butyl)-1,3,5,6- tetrahydroxy xanthone. Sapwood contains erythrodiol acetate, γ-sitosterol, friedelin and friedelan-3-ol in addition to the xanthones reported from heartwood. Stem bark contains tannins (9%), inophyllic acid, inophylloidic acid, friedelin, β-sitosterol and β-amyrin. The leaves are found to yield biflavones (amentoflavone), friedelin, canophyllal, canophyllol, canophyllic acid and piscicidal inophyllum A, B & D and *cis*- and *trans*-inophyllolides. Petals yield procyanidin and stamens, quercetin and myricetin (Goh & Jantan, 1991).

The oil and seed paste are useful in painful joints, rheumatism and gout. Calophyllolide, inophyllolide and calophyllic acid are reputed remedies of leprosy and, therefore, the oil is intramuscularly injected in leprosy patients. The other uses of oil include its application against veneral diseases, scabies, cutaneous diseases and ringworms. Calophyllolide, isolated from defatted seed meal, possesses anti-coagulant, antiarrhythmic, coronary dilator and CNS depressant activity. Leaves are inhaled for migranine and vertigo. Stem bark is purgative. The gum exudate from bark is emetic, purgative and is used for ulcers. Flower decoction is used for syphilis and eczema. Flowers and stamens are substituted for the Ayurvedic drug *Nagkesar* (*Mesua ferrea* Linn.).

3.6 Triterpenoids

The term triterpenoid refers to a heterogeneous collection of biochemical substances, which are believed to be derived from the C-30 acyclic compound squalene by ring closures and substitutions. Apart from a few acyclic members, the majority is represented by tetra or pentacyclic compounds. One may frequently encounter alcohols, aldehydes, ketones or carboxylic acids in them. Many are colourless crystalline compounds with high melting points. Optical isomerism is a common phenomenon exhibited by these natural products. **Tetracylic triterpenes** are recognized by a cyclopentanoperhydrophenanthrene nucleus and an 8-carbon side chain at C_{17}. The resemblance with the steroids is apparent and erroneously prompted some authors to group them as sterols and name them accordingly, e.g., lanosterol. They are abundant in laticiferous plants. Cucurbitacins, a group of bitter principles (12 have been identified occurring free or as glycosides), represent a physiologically active set of compounds in this series.

Pentacyclic triterpenes are widely distributed in higher plants, occurring free or as glycosides (saponins). The aglycones, found in the excretions and cuticle, have a protective as well as water-proofing function, e.g. β-amyrin. Invariably all the members are oxygenated at the C_3 position. The bitterness exhibited by some of these members are notable, e.g. limonoids and quassinoids. Gymnemic acids, which can destroy the ability to taste sweet substance form another interesting group.

Steroids

Steroids also possess the cyclopentanoperhydrophenanthrene nucleus and 8-carbon side chain at C_{17}. But they differ from the tetracyclic triterpenes by the reduction of methyl groups to 2, i.e. only at positions 10 and 13. These compounds, once believed to be of animal origin, are extensively located in higher plants now, where they occur free, as esters or as other derivatives. The typical "animal sterols" like cholesterol and hormones like estrones and ecdysones are reported recently from many plants, but their role in plants is still very ambiguous. Ergosterol is confined to fungi whereas sitosterol, stigmasterol and campesterol are widely distributed in the plant kingdom. More than 100 plant sterols have been identified.

Sterolins and Saponins

Sterolins are sterol glycosides which occur along with free sterols in waxes and unsaponifiable lipid fractions. Unlike saponins, they are non-toxic to fish and do not haemolyse RBC. But the lesser number of sugar moieties—normally only one—make them insoluble in water. They possess high m.p. like sterols but are less soluble in lipid solvents. β-Sitosterol glycoside is the commonest sterolin.

Saponins, on the other hand, are soluble in water and like soaps, produce stable froth when shaken with water. Toxicity to fish and haemolysis of RBC are two distinct properties of this group. The sugar component, usually an oligosaccharide consisting of 2-5 sugar units and one glucuronic acid molecule, is linked to C_3 position of the aglycones. Sapogenins, the aglycones, may be a triterpene like oleanolic acid or a steroid possessing a spiroketal side chain like digitogenin. A series of saponins may have different sugar components but same sapogenin.

Saponins are widespread in higher plants, having been detected in well over 70 plant families. Steroid saponins are frequent in monocot families like Liliaceae, Amaryllidaceae, and Dioscoreaceae, whereas triterpene saponins are predominant in dicots.

3.6a Triterpenes and Steroids

Allamanda cathartica Linn. (Apocynaceae)

Campanilla

A native of S. America, *A. cathartica* is a tall shrub with whorled, obovate to lanceolate leaves, large yellow flowers in cymes and prickly capsules containing many obovate flat, winged seeds. Campanilla is generally grown in gardens for its beautiful flowers.

The leaves contain triterpene esters, plumericin, isoplumericin, plumieride, ursolic acid, β-amyrin and β-sitosterol. The stem also contains most of these compounds. The roots contain an iridoid lactone allamandin, allamandicin, allamdin and triterpene lactones such as fluvoplumeirin, plumericin, isoplumericin and plumieride as well as lupeol and fatty alcohols and fatty acids. The petals yield flavonoids, quercetin and kaempferol.

The leaves are used as cathartics. The extracts of roots and leaves cause hypotension in male cats. The bark is useful as a hydragogue in ascitis. Allamandin exhibits anti-leukemic properties.

Achyranthes aspera Linn. (Amaranthaceae)

Apamargah, Prickly-Chaff Flower

A. aspera is an erect pubescent herb, with opposite, short-petioled, obovate, softly tomentose leaves and terminal spikes containing small greenish-white sessile deflexed flowers. Bracts and bracteoles of the flower are spinescent. Seeds are subcylindric, reddish brown and truncate at the apex. A native of S.E.Asia, this plant is found throughout India as a weed.

All the parts of the plant are found to contain ecdysterone, an insect-moulting hormone. The whole plant yields two alkaloids achyranthine and betaine; leaves contain phenolic acids such as vanillic, syringic and *p*-coumaric acids, roots contain oleanolic acid besides ecdysterone, and the seeds yield protein (22%), carbohydrates (56%) and two saponins, saponin A and B, based on oleanolic acid.

The entire plant is medicinally useful. Young leaves are consumed as a vegetable. The whole plant is much valued for its pungent, diuretic and emmanagogue properties. The decoction is useful in pneumonia, cough and kidney stone. Leaf extract is used to treat eczema, dropsy, for treating cataracts and tetanus. Leaves are also used as a cure for gonorrhoea and leprosy. Root paste is applied to remove opacity of cornea and as a haemostatic and is used in abdominal tumours and as an antifertility agent. Seeds are used against bronchial affections. Achyranthine is reported to dilate blood vessels, lower blood pressure and increase the rate and amplitude of respiration.

Siddha physicians use apamargah as an effective remedy for bronchial asthma (Suresh et al. 1985). Wadhwa et al. (1986) found the plant to be an effective contraceptive in rats and hamsters.

Trianthema portulacastrum Linn. (Aizoaceae)

Syn. *T. monogyna* Linn.

Punarnava

A prostrate succulent mucilaginous herb with fleshy, obliquely opposite leaves. The leaves are in unequal pairs, the larger orbicular-ovate and smaller oblong apiculate. Flowers are axillary, solitary and enclosed and sunken in the membranous enlargement of the petiole of the smaller leaves, forming a pouch. Flowers are apetalous, sepals pinkish white and ovary unicellular. The fruit is in the form of a capsule with two spreading teeth. Concealed by the pouch, it opens by circumscissile dehiscence. Seeds are many, black, reniform. A common weed throughout India, this plant is found in moist places.

The roots contain ecdysterone and an alkaloid, trianthemine ($C_{32}H_{46}O_6N_2$). Leaves are found to posses sesuvin (6,7-diOMe, 3,5,4'-triOH flavone), vanillic acid and *p*-hydroxy benzoic acid (Daniel and Sabnis, 1986).

Roots are considered punarnava (in place of *Boerhavia diffusa*), a rejuvenative drug specific to general weakness, jaundice, oedema, etc. The drug is a laxative, diuretic, expectorant, antipyretic and cardiotonic. *T. monogyna* is found to be analgesic and antiinflammatory. The whole plant is used as a vegetable.

Vernonia cinerea Less. (Asteraceae)

Syn. *Conyza cinerea* Linn.

Sahadevi

Sahadevi is an erect hispid annual with leaves variable from broadly elliptic to spathulate, margin wavy/crenate, pinkish violet homogamous heads in lax terminal corymbose panicles, flowers are in the form of disc florets, pappus of 2 whorls of white hairs and 5-angled achenes bearing appressed white hairs. This plant is found as a weed throughout India and Africa.

The herb contains β-amyrin, lupeol, β-sitosterol, stigmasterol, α-spinasterol and phenols.

Sahadevi is a reputed drug for all types of fevers. The entire plant is febrifuge, diaphoretic and can calm the mind. It is also useful in leucorrhoea, excessive bleeding, skin diseases, and conjunctivitis. Flowers are used in conjunctivitis and rheumatism. Seeds are anthelmintic and alexipharmic and for skin diseases like leucoderma and psoriasis.

Gossypium arboreum Linn. (Malvaceae)

Karpasa, Cotton Tree

G. arboreum is an erect arborescent shrub with purple hairy branches, palmately 5-7 lobed (with an extra tooth in the sinus) lower surface punctate with black dots, long-petioled bristle-tipped leaves, large purple (or yellow with purple base) flowers having 3 large, foliaceous, ovate-cordate bracteoles (flowers with monadelphous stamens and reniform anthers) and a loculicidal capsule containing many seeds covered by cotton fibres. A native of India and Africa, Karpasa is cultivated in gardens and temples and not as a field crop or seen wild.

The seed is considered an aphrodisiac, galactogogue and an effective medicine for anaemia and genito-urinary diseases. The leaves are diuretic, used in mental disorders and skin diseases. Flowers are blood purifiers and heal ulcers due to leprosy. Roots are used to bring down fever.

Gossypium herbaceum Linn.

Kapas, Cotton Plant

G. herbaceum is an erect shrub differing from *G.arboreum* in having yellow flowers or yellow with purple base, becoming reddish after maturity. This is the cultivated cotton plant.

Cottonseed is known for the oil (30%), protein and the dimeric naphthalene derivative, gossypol. Gossypol is present in root bark along with β-sitosterol and α-amyrin.

G. herbaceum is used in place of *G. arboreum* and possesses most of the medicinal properties of the latter. Root bark is emmenagogue and useful in dysmenorrhoea. Flowers are useful in uterine discharges. Gossypol is used in treating endometriosis and uterine bleeding. Gossypol is considered a safe antifertility drug for males.

Helicteres isora Linn. (Sterculiaceae)

Mrigashinga/Maraphali, East Indian Screw Tree

H. isora is a small stellately pubescent tree with oblong-obovate cordate acuminate stellately hairy leaves, red fading to lead-colored flowers in axillary clusters of 2-6 (flowers with 2-lipped calyx, unequal reflexed petals and a deflexed staminal column fused with gynophore) and follicles (5) spirally twisted in the form of a screw containing many small tubercled seeds. This medicinal plant is common all over India and nearby countries.

The plant contains a 4-quinolone alkaloid malatyamine. Root and stem yield β-sitosterol, betulic acid, oleanolic acid and daucosterol. Roots also contain a triterpenoid, isorin and 3β-27-acetoxy-lup-20(29)-en-28-oic methyl ester. Fruits contain α- and β-amyrins, lupeol and its acetate, friedelin, epifriedelinol, bauerenol acetate and taraxeron.

The roots of this plant are used in cough and asthma. The leaf paste is effective against eczema and skin ailments. Pods are antidysenteric and used for colic, flatulence and stomach ache.

Hemidesmus indicus R.Br. (Asclepiadaceae)

Syn. *Periploca indica* Linn.

Anantamool/Sariva, Indian Sarasaparilla

This is a twining undershrub with a woody rootstock, many slender laticiferous branches thickened at the nodes, leaves varying from elliptic-oblong to linear-lanceolate, and often variegated with white above, small greenish yellow/purple flowers crowded in subsessile axillary cymes in the opposite axils and a pair of divaricate long slender follicles containing flattened oblong comose seeds. A native of India, this plant is common throughout.

The roots, which are used as the drug, contain an essential oil consisting of *p*-methoxy salicylic aldehyde; coumarino-lignoids, hemidesminine ($C_{23}H_{22}O_8$) and hemidesmin 1 & 2 and steroids such as hemidesmol, hemidosterol, lupeol octacosanate and β-amyrin acetate and coumarin. Stem contains pregnane glycosides hemidine, hemidescine, emidine, indicine, a tritepene lactone (3-keto-lup-12-ene 21(28)-olide), a lupanone, Δ^{12} dehydrolupanyl-3β-acetate, besides Δ^{12} dehydrolupeol acetate, lupeol acetate, sitosterol and several hydroxymethoxy benzaldehydes (Mandal et al. 1991; Gupta et al. 1992; Chandra et al. 1994).

Sariva is one of the rejuvenating drugs used for its cooling and blood purifying qualities. It is an alterative, aphrodisiac, refrigerant, diuretic and tonic. It is also useful in chronic rheumatism, anaemia, dyspepsia, leucorrhoea, uterine haemorrhage and skin diseases as well as abdominal tumours and as an antilithic.

Ichnocarpus frutescens R.Br. (Apocynaceae)

Sariva, Ichnocarpus

I. frutescens is an extensively branched woody laticiferous twiner with rusty pubescent young branches, elliptic-oblong leaves, small greenish white flowers in axillary or terminal rusty pubescent trichotomous pedunculate cymes and a pair of cylindrical follicles enclosing black linear seeds crowned by a tuft of white hairs. This plant is distributed throughout India, Malaya and Australia.

Leaves contain flavones apigenin and luteolin, glycoflavones, vitexin and isovitexin, proanthocyanidins and phenolic acids, vanillic, syringic and sinapic acids (Daniel and Sabnis, 1978). Stem contains friedelin, friedelinol, lupeol acetate, Δ^{12} dehydrolupeol, Δ^{12} dehydrolupanyl, oleanolic acid, β-sitosterol and α-amyrin glycoside (Verma and Gupta, 1988).

This plant is used as Sariva (in place of *Hemidesmus*). Various plant parts are used in night blindness, bleeding of gums, enlargement of spleen, atrophy, smallpox, ulcer, dysentery, cough and asthma. It is also used in abdominal and glandular tumours.

Gymnema sylvestre R.Br. (Asclepiadaceae)

Madhunashi, Gymnema

G. sylvestre is a large, much-branched woody twiner, with opposite pubescent ovate-lanceolate leaves, small yellow flowers in sessile or pedunculate cymes (flowers with staminal corona and a gynostemium) and an etaerio of lanceolate beaked follicles (one follicle often suppressed) enclosing flat ovoid-oblong seeds having a thin broad marginal wing. Common in Western ghats, Gymnema is a native of tropical Africa.

The leaves contain gymnemic acids (with gymnemic acid being the main component), 3β-glucuronides of acylated gymnemagenins (hexa hydroxy olean-12-one), gymnestrogenin, conduritol A, nonacosane, choline, betaine and an alkaloid gymnamine, and tritriacontane. Also present are dammerane type saponins gymnemasides l-Vlll and gypenosides XXVll, XXXVlll, LV and LXlll. The phenolics present are flavonols (quercetin and 3'OMe quercetin) and phenolic acids (vanillic, syringic, *p*-hydroxy benzoic, protocatechuic, ferulic and sinapic acids) (Daniel and Sabnis, 1982).

The leaves and stems of this climber are hypoglycemic and, therefore, used in treating diabetes. They are diuretic and stimulate cardiovascular system. The drug is useful in skeletal fractures and exerts a purgative action.

Holoptelia integrifolia Planch (Ulmaceae)

Putikaranjah, Holoptelia

H. integrifolia is a large glabrous deciduous monoecious tree with elliptic glabrous leaves, male and female flowers mixed in short racemes from the scars of fallen leaves and a orbicular samaroid fruit 2-3 cm in dia. having reticulately veined wings. A native of China, Holoptelia can be seen in all deciduous forests.

The bark yields β-sitosterol, friedelin, friedelan-3β-ol and triterpenoidal fatty acid esters, holoptelin A (*epi*-friedelinol palmitate) and B(*epi*-friedelinol stearate). Heartwood contains dihydroxy olean-12-en-28-oic acid, and the leaves contain β-sitosterol, β-amyrin and hexa- and octacosanols.

Both the bark and leaves are used (normally in combination with Karanjah; *Pongamia pinnata*) as laxative and anthelmintic. They are also used for oedema, leprosy, skin diseases and diabetes. Branches are used as fish poison, stem bark and seeds applied to cure ringworm. Stem bark is also used in scabies and rheumatism. The drug exhibits lipolytic action and, therefore, helps reduce obesity.

Commiphora spp. (Burseraceae)

Myrrh

Myrrh is an oleo-gum-resin obtained from various species of *Commiphora*; growing in Africa, Arabia and Abyssinia. Apart from *C. myrrha* Engl., which forms the principal source, *C. abyssinica* Engl., *C. schimperi* Engl. And *C. erythraea* Arn. also yield this valuable resin.

The resin, occurring in the schizolysigenous cavities of phloem, exudes naturally from the spontaneous cracks or crevices developed in the bark or can be obtained by making incisions. The yellowish white liquid oozing out gradually solidifies to form reddish brown irregular masses, which is then collected.

Myrrh contains 5-10% volatile oil, 25-40% resin, 55-60% gum and some bitter principles as impurities. The yellow-green volatile oil consists of eugenol, cuminaldehyde, mono- and some sesquiterpenes. The resin—the components of which are not fully characterized—contains α-, β- and γ-commiphoric acids; α- and β-heerabomyrrholic acids and some phenolic compounds such as protocatechuic acid and pyrocatechin. The gum portion is composed of about 20% protein and 60% carbohydrates, which on hydrolysis yield galactose, arabinose and glucuronic acid as principal components.

The oldest and most valuable gum-resin, myrrh is used as an incense, in perfumes and embalming. In medicine it is employed as a tonic, stimulant and antiseptic.

Commiphora wightii Arnott — (Burseraceae)

Guggul

Guggul is the oleo-gum-resin obtained from *C. wightii*, a native of India. The plant is a small tree, branched from the base with knotty divaricate glandular pubscent branches ending in sharp spines. Leaves are 1-3 foliolate with shining rhomboid ovate leaflets. The red-coloured drupe contains two woody pyrenes.

The resin, which is exuded naturally or by incisions, forms pale yellow or reddish brown vermiform crystals and possesses a bitter aromatic taste. It contains about 0.37% essential oil, consisting chiefly of myrcene, dimyrcene and polymyrcene; sterols like guggulusterol I, II, & III, cholesterol, guggulusterones Z & E, diterpenes such as α-camphorene, cembrene and mukulol and long chain aliphatic tetrols and sesamin. Muscanone, 3-O- (1″, 8″, 14″ trimethyl hexadecanyl) naringenin is an antifungal flavone isolated from this plant along with naringenin (Fatope et al. 2003).

Guggul is used in treating rheumatism, obesity, and neurological and urinary disorders. Guggulusterols are found to be anti-inflammatory and hypocholesteremic.

Aerva lanata Juss. (Amaranthaceae)

Bhadra, Aerva

A. lanata is a much-branched woody herb with simple alternate ovate/orbicular leaves, minute sessile white flowers in axillary spicate clusters and reniform seeds enclosed in greenish compressed utricles. Aerva is a common weed throughout India.

The plant is found to contain six alkaloids, canthin-6-one, 10-OMe canthin-6-one, 10-OH canthin-6-one, β-carboline-1-propionic acid and its 6-OMe deriv. and glucooxycanthin-6-one. Also present are β-sitosterol, α-amyrin, betulin, four coumaroyl glycosides, phenolic acids such as vanillic and syringic acids and saponins. Roots yield aervin, methylaervin and aervoside (Zepesochnaya, et al. 1991).

The plant is anthelmintic, demulcent and used against kidney and urinary stones and diabetes. It is also used in malaria and skin diseases and as an expectorant and for indigestion and wounds. The roots are often preferred, though the whole plant is medicinal. *A lanata* is one of the 10 auspicious herbs that constitute the group *Dasapushpam* ('ten flowers').

Drypetes roxburghii Hurus. (Euphorbiaceae)

Syn. *Putranjiva roxburghii* Wall.

Putranjiva

This is an evergreen/dioecious tree (upto 20m ht.) with pendent branches, dark green, shinning elliptic-oblong serrulate penninerved leaves, shortly pedicellate male flowers crowded in globose axillary clusters, female flowers 1-3 in axils and ellipsoid white tomentose drupes containing a single hard seed enclosed in a hard endocarp. Putranjiva is common in tropical India.

The stem bark yields triterpenoids friedlin, friedelanol, roxburgholone, putranjivadione, putranjivanol, putranijic acid and putrolic acid. Leaves contain β-amyrin, putrone, putrol, putranjivic acid, methyl putranjivate, stigmasterol, roxburghonic acid, saponins A-D and biflavones. Seeds yield fatty oil and an essential oil containing isothiocyanate produced from glucosinolates glucopatranjivin, glucocochlearin, glucojiaputin and glucolemin. Seed coat yields triterpenoid saponin putrainjivoside (1.3%), β-sitosterol and its glycosides and saponins such as putranosides A-D. The fruit pulp contains mannitol, a saponin and an alkaloid.

Leaves and stones of the fruits are used in rheumatism. Leaves are also used externally to swollen joints and inflamed areas. All parts of the plant are used for colds and fevers.

Euphorbia hirta Linn. (Euphorbiaceae)

Syn. *E. pilulifera* Linn.

Dugdhika

E. hirta is a laticiferous erect herb with long yellowish crisped hairs, opposite obliquely oblong-lanceolate (base unequal sided) leaves, pedunculate umbellate clusters of cyathia and pubescent capsules containing 3 reddish brown trigonous seeds. Dugdhika is a very common weed found all over India.

The whole plant yields cycloartenol, friedlin, taraxerol, aphyldienol, ingenol triacetate, euphorbol hexacosanate, β-amyrin acetate, tinyatoxin, β-sitosterol, cycloartenol, choline, shikimic acid and inositol. Also present are quercetin, rutin, an alkaloid xanthorhamnine and dimeric tannins euphorbins A, B & C (composed of geraniin and pentagalloyl glucose). The latex contains diesters of 12-deoxy-4β-phorbol (Okuda et al. 1993).

This plant is used as a diuretic, aphrodisiac, and galactogogue, for skin diseases, asthma and other respiratory diseases. It is also used as an antidysenteric, vermifuge and in diseases of the urinogenitory tract. It also exhibited anticancer properties in the sense that the aqueous extract reduced release of prostaglandins l_2, E_2 and D_2 and had an inhibitory effect on platelet aggregation and carrageenan-induced rat paw oedema.

Euphorbia thymifolia Linn.

Syn. *E. prostrata* Ait.

Dugdhika

This plant is also used as Dugdhika, but it is further administered in ophthalmia and other eye troubles, atrophy, dysentery and breast pain. This is considered as an insecticide, a blood purifier and cures spermatorrhoea.

The entire plant yields deoxyphorbol acetate derivatives, epitaraxerol hexacosanol, euphorbol, 24-methylene cycloartenol, taraxasterol, tirucallol and quercetin.

Euphorbia nivulia Buch. - Ham. (Euphorbiaceae)

Syn. *E. neriifolia* Linn.

E. nivulia is a large fleshy, armed erect shrub with pairs of sharp stipular spines arising from conical truncate spirally arranged tubercles. Leaves, seen towards the end of branches, are fleshy, alternate, obovate-oblong or spathulate and tapering towards the base. Cyathia are in twin cymes from above the leaf sears. The capsule contains 4-angular smooth seeds. This plant is common in wastelands.

The latex contains euphol and nerifloiol, while the bark yields deoxyphorbol acetate, euphol, euphorbol hexacosanate, 24-methylene cycloartenol and glycosides of pelargonidin and tulipanin. Leaves and stem are found to possess friedelan-3α-ol, taraxerol and glut -5 (10)-en-1-one, whereas the roots contain 24Me-25-ene cycloartenol, cycloartenol, ingenol triacetate, euphorbol and deoxyphorbol acetate.

The stem, with a milky latex, is used as a purgative, expectorant and digestive and cures liver and spleen enlargement, asthma, fever, leprosy and rheumatism. The juice is applied externally to remove warts and similar structures, ulcers, scabies, etc.

Clerodendrum indicum Kuntze (Verbenaceae)

Syn. *C. siphonanthus* C.B. Clarke

Bharangi

C. indicum is a woody shrub with an obtusely 4-angled stem, simple lanceolate serrate thick leaves in whorls of three white flowers in a terminal panicle and purple drupe containing 4 seeds, party enclosed in calyx. Bharangi is a common plant throughout India.

The stem contains β-sitosterol and cholesta-5,22,25-triene-3β-ol. Leaves contain scutellarein and hispidulin, whereas the flowers yield β-sitosterol.

Root, which is the official drug, is antispasmodic, expectorant, febrifuge and used for epilepsy and dropsy. The leaves are suppurative, while the fruit decoction is said to have antifertility factors.

C. serratum Moon. is used as bharangi in some plants of India.

Ficus benghalensis Linn. (Moraceae)

Nyagrodhah/Vad, Banyan Tree

F. benghalensis is a very large laticiferous tree with numerous aerial roots from the branches which take root on the ground and thus extend the growth of the tree infinitely. Leaves are large, broadly elliptic, glabrous above and minutely pubescent beneath. Figs are monoecious, axillary depressed globose 2 cm in diam, brick red when ripe with 3 broad rounded basal bracts. Male flowers are numerous near the mouth, female flowers at the base of receptacle and neuter flowers in the middle. Achenes are globose, ellipsoid. A native of India, banyan tree is cultivated or found wild through out India.

The stem bark contains bengalenoside, procyanidin, β-sitosterol, *meso*-inositol and ketones. Leaves yield friedelin and β-sitosterol, quercetin and rutin. Heartwood contains ψ-taraxasterol.

All the parts of the tree are used in medicine, especially for skin diseases. The bark is useful in menorrhagia, leucorrhoea and other vaginal disorders and is found to cure diarrhoea and dysentery. Bengalenoside is found to be hypoglycemic. Latex and young buds are used in urinogenital diseases. The roots are also useful in haemoptysis, menorrhagia and ulcers. Young roots are found to cure pimples.

Ficus carica Linn.

Anjir, Fig

This is a middle-sized laticiferous deciduous tree with cordate, 3-5 nerved, dentate and deeply palmately lobed leaves. Figs are large and edible.

The plant yields umbelliferone and scopoletin. The leaves contain psoralens, bergaptene, xanthotoxin, xanthotoxol and marmesin. Flavones, terpenoids and fiscusogenin are the other compounds reported. Latex contains ficin and peptides. Roots yield guaiazulene.

The fruit is useful in anaemia. Latex is anthelmintic. Latex exhibited significant inhibition of the binding of 3H-benzo α-pyrene, a known chemical carcinogen. Three peptides are also reported from latex which exhibited inhibitory action against Angiotensin-1-converting enzyme (ACE). The fruit extract contains an anticancer compound effective against Ehrlich sarcoma.

Ficus racemosa Linn.

Syn. *F. glomerata* Roxb.

Udumber

This is an evergreen laticifer with elliptic lanceolate leaves tapering to a blunt pointed apex, ovate-lanceolate pubescent stipules and red dense tomentose figs borne on short leafless warted short branches of a few inches long, subglobose or pyriform with 3 basal bracts and monoecious. Udumber is found throughout India.

Bark contains ceryl behenate, lupeol, α-amyrin and β-sitosterol and gluanol acetate. Leaves yield β-amyrin, β-sitosterol and gluanol acetate. The fruit also contains gluanol acetate along with lupeol acetate, β-sitosterol and taraxasterol.

The bark of this tree is very useful in healing ulcers, skin diseases and vaginal diseases. It is highly effective in threatened abortions, gonorrhoea, menorrhagia and leucorrhoea. Leaves are hypotensive and cardiac depressant. The fruits are used to treat diarrhoea, dysentery, dyspepsia, hemorrhage and menorrhagia.

Ficus religiosa Linn.

Aswattah

F. religiosa is a large glabrous laticiferous tree with long-petioled, broadly ovate rotund leaf having a long lanceolate cuspidate tip and monoecious paired dark purple figs. Aswattah is a highly venerated tree found throughout India.

The bark contains bergenin, lupin-3-one, methyl oleanolate, lanosterol, n-octacosanol, β-sitosterol, stigmasterol and caffeic acid (Swami et al. 1989).

The bark is alterative, haemostatic, laxative and is used in diabetes, diarrhoea, leucorrhoea, menorrhagia, nervous disorders and vaginal diseases. It is also an excellent remedy for skin diseases and is anthelmintic. Leaves are also used in skin diseases. Seeds have the power to sterilize women.

Cissus quadrangula Linn. (Vitaceae)

Syn. *Vitis quadrangula* Wall.

Asthisandana/Vajravalli, Bonesetter

C. quadrangula is a succulent perennial climber with quadrangular jointed cactus-like stems having 4 wings, leaf-opposed tendrils, aerial roots from nodes, cordate fleshy 3-7 lobed/entire leaves (leaves fall off in older stems), small greenish-white flowers in short umbellate cymes and globose, pea-sized berries enclosing solitary seeds. This is a native of S.E. Asia.

The stem contains protein (12%), carbohydrates (35%), carbonates of sodium, potassium, magnesium and calcium, potassium tartarate and triterpenoids. The triterpenoids present are onocer-7-ene-3α, 21β diol, onocer-7-ene-3β, 21α-diol, 7-oxo-onocer-8-ene-3β-diol, alkanes and two steroids designated I and II besides taraxasterol, friedelan-3-one, β-sitosterol, δ-amyrin and δ-amyrone (Gupta & Varma, 1991).

The stem is useful in piles, asthma, irregular menstrual periods and muscular pains. But the entire plant is excellent in healing bone fracture because of the two unidentified anabolic steroids I and II, which facilitate early regeneration of all connective tissues involved in healing and quicker mineralization of the callus. The aqueous extract of the plant is administered topically or as intramuscularly injection.

Withania somnifera Dunal (Solanaceae)

Ashwagandha, Indian Ginseng

W. somnifera is a green herbaceous perennial with long hard tuberous roots, dichotomously branched stem, ovate leaves and small sessile greenish-yellow flowers crowded in the axils of leaves. The small berry is enclosed in a green persistent inflated calyx. This medicinal plant is widely distributed in north western India.

The roots, which form the drug, exhibit a great chemical diversity. The majority of constituents are a group of steroidal lactones possessing an ergostane skeleton, the withanolides. The known withanolides are more than 20 and are withanone, withaferin A, withanolides I, II, III, A, B, C, D, E F, G, H, I, J, K, L, M, WS-1, P and S, withasomnidienone and withanolide C. About a dozen biochemically heterogeneous alkaloids are reported from this drug. The major compounds are tropane, pseudotropane (tropanes), hygrine (pyrrolidine), isopelletierine (piperidine), anaferine (two piperidine moieties), withasomnine (pyrazole) and anahygrine (having one pyrrolidine and one piperidine moieties). Also present is a large amount of starch (Bakuni and Sudha, 1995; Bessale and Lavic, 1992).

Ashwagandha is a rejuvenating drug used for rheumatism, hypertension, and as a tonic. It shows antitumour, bradycardic, respiratory stimulant and antispasmodic properties. Withanolides also show antitumour, antiarthritic and immunosuppressive properties (Sudhir et al. 1986).

Citrullus colocynthis Schard. (Cucurbitaceae)

Syn. *Colocynthis vulgaris* Schard.

Indravaruni, Colocynth

C. colocynthis is a monoecious perennial with scabrid angular stems, bifid tendrils, triangular deeply 3- lobed (lobes pinnately divided) leaves, yellow unisexual flowers solitary in the axils of leaves and green fruits (pepo) mottled with yellow blotches containing spongy bitter pulp and numerous white seeds. Colocynth is abundant in northwestern plains of India on sand dunes.

The fruit contains a number of bitter principles generally known as cucurbitacins consisting of α-elaterin (cucurbitacin E), elatericin B (cucurbitacin I), dihydroelatericin B (cucurbitacin L), citrullin, citrullene, citrullic acid and citrullol. Also present are sugars (glucose), pectin, hydrocarbons such as hentriacontane, a volatile oil consisting of citronellal, methyl eugenol and methyl heptenone, choline, alkaloids, lanosterol and β-sitosterol. Seeds yield a bitter oil consisting of linoleic (60%), oleic (20%) and palmitic (10%) acids and a triterpene, citrullonol (Hatam et al., 1989).

The dried pulp, containing seeds (62%) and rind (23%) is the drug *colocynth*. It is a hydragogue, cathartic and drastic purgative and therefore used along with carminatives. It is also useful in treating constipation, dropsy, and fevers and useful as a vermifuge and emmenagogue. The purgative property is attributed to α-elaterin glucoside.

3.6b Limonoids and Quassinoids

Limonoids and Quassinoids are the compounds formed by the loss of a few carbon atoms from triterpenoids and thus known as nortriterpenoids. All these compounds are unique to the order Rutales *sensu lato*. Limonoids are tetranortriterpenoids formed by the conversion of the 8-carbon side chain at C-17 to a furan ring. But some of the members of this group, like turreanthin, retain an intact C_{30} triterpenoid skeleton. Azadirone, nimbin, trichilin, etc., are C_{26} limonoids. Almost all these compounds are bitter principles. Azadirachtin is a complex member of this group showing marked antifeedant activity. Limonoids are subdivided based on the cleavage of D ring (gedunin), C ring (nimbin), B ring (andirobin), A ring (tricoccin S_{22}), both A & B rings (dregianin), both A and D rings (limonin) or with all 4 rings intact (azadirone). All these limonoids are restricted to the Meliaceae and Rutaceae. The Cneoraceae, a related small family, is found to contain C_{25}, pentanortriterpenoids (cneorin B). About 300 limonoids are known (Conolly, 1983; Dreyer, 1983).

Quassinoids are the characteristic bitter principles of the Simaroubaceae. They are highly degraded triterpenoids, which lost the 8-carbon side chain together. They possess C_{30} skeleton (glaucarubolone), C_{19} (samaderine C), C_{18} (laurycolactone A) or rarely C_{25} (simarinolide). About 100 quassinoids are known today (Polonsky, 1983).

Some of the quassinoids, like isobruceine A possess marked antileukaemic activity. Other biological activities include: (1) antiviral activity (castelanone), (2) antimalarial (glaucarubinone), and (3) antifeedant and insecticidal properties (most of limonoids and quassinoids).

Picrasma excelsa Planch. (Simaroubaceae)

Quassia

A native of West Indies, this plant is a tall polygamous shrub with very large compound leaves having numerous ovate unequally sided leaflets.

The fine shavings of the wood, the Quassia of commerce, contain a mixture of bitter terpenoids like the lactone quassin (which form the chief constituent), neo-quassin (the hemiacetal of quassin) and picrasmin (*iso*-quassin).

Quassia is a bitter tonic, an insecticide and anthelmintic.

Quassia amara L., a native of Brazil, contains similar principles and is used in place of *Picrasma*. Recently, it has been observed that quassinoids possess antileukaemic properties. One such compound under study is Bruceantin isolated from *Brucea antidysenterica* Lam. (Simaroubaceae).

Ailanthus excelsa Roxb. (Simaroubaceae)

Ardusa, Tree of heaven

A. excelsa is a large majestic, deciduous tree with a tall cylindrical bole, pinnate leaves (upto 90cm long, having 8-14 pairs of unequal sided leaflets), small yellowish flowers in panicles and one-seeded samaroid fruits. This tree is very fast growing and is introduced all over India.

The stem bark contains quassinoids such as glaucarubin and excelsin, ailanthic acid, 2,6-dimethoxybenzoquinone, β-sitosterol, malanthin, triacontane and hexatriacontane. The root bark contains quassinoids such as ailanthione, glaucarubinone, glaucarubol-15-isovalerate and 13,18-dehydroglaucarubol-isovalerate. Glaucarubonone is anthelmintic and possesses anti-spasmodic and expectorant properties. It is used for asthma, bronchitis, dysentery, dyspepsia and earache. The bark is a substitute of kurchi (*Holarrhena antidysenterica*). All the quassinoids of root bark are found to possess substantial anti-tumour and cytotoxic activities against the P-388 lymphocyclic leukaemia and KB test systems, respectively. The leaves are used as an adulterant for *Adhatoda zeylanica*.

Azadirachta indica A. Juss. (Meliaceae)

Nimba, Neem

Neem is a large evergreen tree with imparipinnate leaves having ovate lanceolate, serrate unequal-sided leaflets, small white scented flowers in axillary panicles and green drupes containing a large seed. Neem is found all over India, cultivated or naturally grown.

All the parts of the plant contain limonoid bitter principles. The stem bark yields nimbin, nimbidin, β-sitosterol, tannins (12%) and a gum; the root bark also contains nimbin and nimbidin; the leaves contain nimbin, nimbinene, 6-desacetylnimbinene, nimbandiol, nimbolide, quercetin and β-sitosterol; and the flowers yield β-sitosterol (along with its glucoside), quercetin, kaempferol, thioamylalcohol and benzyl alcohol and a volatile oil containing sesquiterpenes margosene and azadirachtene. The fruits are found to contain gedunin, azadirachtin, 7-deacetoxy-7α-hydroxy gedunin, azadiradione, azadiradone, 17-β-hydroxy azadiradione, 17-*epi* azadiradione and nimbiol. Neem seeds yield oil and a number of bitter principles both in oil and outside. The oil consists of oleic (56%), palmitic (16%), stearic (14%) and linoleic (9%) acids and 0.8% unsaponifiable matter containing nimbidin, nimbin, nimbinin, meliantriol as major components along with gedunin, meldenin, desacetylgedunin, salannin, azadirone, epoxyazadirodione, vepinin, nimbinene, 6-O-acetyl nimbandiol, etc.

The extracts of neem are found to be antimicrobial, anthelmintic and useful against fever especially of malaria. Since all parts are bitter, they are used as anti-diabetic drugs also. The leaves are useful in skin diseases. The fruits also are used as tonics, antiperiodic, purgative, in treating urinary diseases and piles. The oil is a remedy for some chronic skin diseases and ulcers. It is applied externally for rheumatism and leprosy. The flowers are used in cases of atonic dyspepsia and general debility. Nimbidin is an effective drug in acute and chronic inflammation as well as psoriasis.

All the bitter principles of neem are insecticidal and, therefore, the leaves, seed oil, etc., are used thus.

Brucea javanica Merril (Simaroubaceae)

Syn. *B. amarissima* Desv.,
B. antidysenterica Merril

Ya-Danzi, Brucea

Brucea, a native of Ethiopia, is a thickly pubescent dioecious evergreen shrub with imparipinnate leaves (leaflets six pairs + one), small purplish flowers borne in interrupted glomerate spikes and small oval black drupes.

The stem bark contains quassinoids such as bruceantin, bruceantinol, dihydrobruceantinol, bruceine B, bruceanols D, E, F, G & H, yadanzigan, yadanziolide A. and brucicanthinoside C (a glycoside), apotirucallane type triterpenoids bruceajavanin A, dihydrobruceajavanin A and bruceajavanin B and an alkaloidal glycoside of β-carboline, bruacecanthinoide. Also present are β-carboline alkaloids like canthin-6-one, 11-OH canthin-6-one, 11-OH, 1-OMe canthin-6-one and canthin-6-one-3-N-oxide beside brusatol, emodin, chrysophanin, chrysophanol, ethyl gallate and β-sitosterol. The leaves yield a volatile oil of dillapiole (22%), thymol (17%), apiole, α-cedrene, carvone, etc. The fruits of Ya-Danzi contain an alkaloid 4-ethoxycarbonyl-2-quinolone, quassinoid bruceoside C, quercetin, luteolin and vanillic acid. Seeds yield a fatty oil.

The plant, especially the fruit, is traditionally used in China and Indonesia in the treatment of malaria, amoebic dysentery, cancer and parasitic diseases. Bruceajavanin A, dihydrobruceajavanin A and brucicanthinoside inhibited the growth of cultured *Plasmodium falciparum* K1, a chloroquin-resistant strain). Bruceantin is active against B16 melanoma, colon 38 and L1210 and P388 leukaemia. It favoured apoptosis and exerted no toxic side effects (Cuendet and Pezzuto, 2004). Bruceanols are also found to be cytotoxic against many human tumour cell lines (Imamura et al. 1993).

3.6c Saponins

The crude drugs containing saponins are generally used for their detergent properties and some of them, which result in lesser degree of irritation on oral administration, are employed as expectorant and antitussive agents. Saponins are characteristically found to be antimicrobial in nature.

Saponins are widely distributed in the plant kingdom. About 70% of plant families contain them but they are much more common in Araliaceae, Rhamnaceae, Apiaceae, Fabaceae, Caryophyllaceae, Hippocastanaceae, Liliaceae and Dioscoreaceae. Guvanov et al. (1970) found that, of the 1730 plants belonging to 104 families of Central Asia, triterpenoid saponins are seen in 627 spp. and steroidal saponins in 127.

Steroidal saponins are also important as the starting materials for the syntheses of steroid hormones and related medicines.

Polygala senega Linn. (Polygalaceae)

Milkwort or Snakeroot

This herb of 5-20 cm height possesses a thick long taproot, simple estipulate leaves, blue bracteate flowers in racemes and a capsule containing arillate hairy seeds. Milkwort is a native of N.E. America, cultivated extensively in Canada and Japan.

The dry rootstock and roots contain a number of triterpenoid saponins that, on hydrolysis, yield the sapogenins, senegenin (a chlorinated sapogenin), presenegenin, senegenic acid, polygalic acid (hydroxy senegenin) and glucose. 1, 5-Anhydrosorbitol (polygalitol) and sucrose impart a sweet taste to the drug and methyl salicylate the aromaticity. The root also contains a good amount of fats.

Senega is used as a cure for snake bite. It is a known expectorant, emetic and stimulant. The plants are given to cattle to increase the production of milk.

Indian senega consists of the roots of *Glinus oppositifolius* DC which contains oleanolic acid saponins.

Smilax spp. (Liliaceae)

Sarasaparilla

The long thin roots arising from the rootstock of *S. aristolochiaefolia* Miller, *S. regelii* K.& M. or *S. febrifuga* Kunth, constitute the drug Sarasaparilla. All these plants are straggling shrubs native to America and Spain.

The drug contains a number of saponins based on smilagenin and its isomer, sarasapogenin. Sarasaponin, one of the principal saponins, yields sarasapogenin, two molecules of glucose and one molecule of rhamnose on hydrolysis. Smilonin, another saponin, yields smilagenin and as many as five molecules of sugars. The roots contain certain other phytosterols, like β-sitosterol and stigmasterol, resins and a volatile oil.

Sarasaparilla is used in skin diseases, rheumatism and as a flavouring agent. The sapogenins are used as the starting material for the synthesis of cortisone and other steroid hormones.

Panax spp. (Araliaceae)

Ginseng

The sources of ginseng are *P. ginseng* A. Mey (Korean ginseng) in China and *P. quinquefolius* Linn. (American ginseng) in America. *P. ginseng* is a low perennial glabrous herb about 15-45 cm in height with a deep spindle-shaped root and whorls of 3-5 palmate leaves at nodes. Green flowers, numbering 6-20, are borne in an umbel on an elongated peduncle. The fruit is a bright red berry.

The dry roots of 3 to 6-year-old plants form the ginseng of commerce. Korean ginseng contains a number of saponins, ginsenosides R_x (x = o, a, b_1, b_2, b_3, c, d, e, f, 20-gluco-f, g_1 and g_2 - approximately 30). Ginsenoside R_{g-1} is the panaxoside reported earlier. All these glycosides (except R_o) are based on sapogenins 20-S-protopanaxodiol and 20-S-protopanaxotriol, dammarane tetracyclic triterpenes. Ginsenoside R_o possesses oleanolic acid as the sapogenin. In American ginseng, ginsenoside R_{g-1} is absent and R_{b-1} is dominant. Other constituents of ginseng are: (1) essential oil containing polyacetylenes and sesquiterpenes; (2) polysaccharides; (3) peptidoglycans such as panacene as also peptides; (4) steroids like panaxatriol, panaxadiol and protopanaxadiol; and (4) choline, vitamins of B-group, C & E, fatty acids, carbohydrates, amino acids (of which a majority are essential amino acids, and a strong proportion of arginine), minerals like germanium and phenolic compounds such as salicylic and vanillic acids.

Widely used (especially in the Chinese system) as a tonic, stimulant, diuretic and carminative, ginseng is employed in anaemia, insomnia and as an aphrodisiac. The pharmacological actions of individual ginsenosides are sometimes found to work in opposition. For example, of the two main ginsenosides R_{b-1} and R_{g-1}, the former suppresses the central nervous system while the latter stimulates the same. These opposing actions may contribute to the adaptogenic properties of ginseng and its proposed ability to balance body functions. Panacene is hypoglycemic and another peptide present is found to be insulinomimetic. Other properties of ginseng include anti-fatigue, vaso-dilation, anxiolytic, anti-depressant, enhancing energy metabolism, stimulating learning, memory and physical capabilities, supporting radioprotection and providing resistance to infection.

Panax pseudoginseng var. *notoginseng* is the San-chi ginseng that originated in China. In this ginsenosides R_{g-1}, R_e and R_{b-1} are dominant. Japanese Chikusetzu-Ginseng is *P. japonicum* C.D. Meyer containing chikusetsusaponins 1, 1a, 1b, 111, IV and V based on oleanolic acid sapogenin.

Eleutherococcus senticosus Maxim. (Araliaceae)

Siberian Ginseng

Siberian ginseng, which is mainly used in Russia, contains saponins (eleutheroside I, K, L & M), steroid glycosides such as eleutheroside A (a glycoside of daucosterol), lignans such as sesamin and eleutheroside D (di-beta-D-glycoside of syiringaresinol), polysaccharides (eleutherane A-G and eleutheroside C), hydroxy coumarins (isofraxidin), phenyalcrylic acid derivatives (eleutheroside B- a glycoside of syringin) and minerals.

In addition to the various species of ginseng, *Rumex hymenosepalus* (Polygonaceae—Wild red desert ginseng), *Pfaffia paniculata* (Amaranthaceae—Brazilian ginseng), *Pseudostellaria heterophylla* (Caryophyllaceae—a substitute of ginseng), *Caulophyllum thalictroides* (Berberidaceae—yellow/blue ginseng), *Triosetum perfoliatum* (Caprifoliaceae—ginseng) and *Codonopsis tangshen* (Campanulaceae) are the other plants used as ginseng in various parts of the world.

Bupleurum falcatum Linn. (Apiaceae)

Bupleurum

A native of Europe and Himalayas, *Bupleurum* is a perennial, much-branched herb, 1 m in height with simple long linear leaves, yellow bracteate flowers in compound umbels and a cremocarp consisting of 2-ridged mericarps.

The roots are found to yield a number of saponins—saikosaponins A, C, & D being the principal ones. The sapogenins are of oleanane type, e.g. Saikogenin F. (Kubota et al. 1969).

The roots of Bupleurum are used as anti-inflammatory, diaphoretic and often prescribed for liver troubles.

Platycodon grandiflorum DC. (Campanulaceae)

Platycodon

This erect perennial, 1 m in height, bears small, ovate-lanceolate leaves (glaucous blue on the lower surface), solitary blue/white flowers on long peduncles and 5-lobed capsules. The plant is a native of E. Asia.

Of the 8 or more saponins—platicodins A-H—known from the roots, platicodin-D forms the major one. The crude saponin extract on hydrolysis yields platicodigenin as the main sapogenin and polygalacic acid (Akiyama et al. 1972) as well as platycodigenic acids as minor sapogenins (Kubota et al. 1969).

The entire saponin extract is employed as an expectorant and an antitussive agent.

Some of the other important Chinese medicinal plants containing saponins are the following:

Polygala tenuifolia Willd. (Polygalaceae)

The roots of *P. tenuifolia*, known as **Yiian-chi** and **Onji** in Chinese and Japanese systems, are used as a sedative and to strengthen the nervous system. This drug contains onjisaponins A-F (onjisaponin A and B are found to be identical with senegin IV and III, respectively, of *P. Senega*).

Akebia quinata DC (Lardizabalaceae)

The stem bark of this plant, known as **Mu-T'ung** (Chinese) and **Mokutsu** (Japanese), is used as an anti-inflammatory agent, diuretic and menses stimulant. The saponins isolated are akeboside St-e, St-b, St-c, St-d, St-f, St-h, St-j and St-k based on hederagenin and oleanolic acid sapogenins. The seeds of this plant also yield saponins based on hederagenin.

Ziziphus jujuba var. *spinosa* Hu. (Rhamnaceae)

Ziziphus

This native of S. Europe and S. E. Asia is a glabrous spiny tree having fascicled branchlets, simple oblong leaves with stipular spines, small yellow flowers in short axillary cymes and a fleshy sub-globose drupe.

The seeds, which form the drug, contains two saponins jujubosides A & B. Both these saponins, on hydrolysis, yield ebelin lactone (Kawai et al. 1974).

In Chinese medicine, the seeds of Ziziphus are recommended for strengthening the nervous system and for insomnia.

Abrus precatorius Linn. (Fabaceae)

Gunja, Crab's eye, Indian Liquorice

A. precatorius is a greenish twiner with imparipinnate leaves having 16-40 oblong leaflets, pale purple flowers in axillary inflorescences, rectangular bulky pods containing ovoid glossy scarlet seeds with a black spot (or black with white spot or uniformly black/white). This plant is a native of India and Malaya.

The leaves, which are sweet in taste, contain up to 10% glycyrrhizin, triterpene glycosides, pinitol and alkaloids such as abrine, hypaphorine, choline and precatorine. The triterpene glycosides are abusosides A, B, and C (which are highly sweet) and three glycosides based on cycloartane-type aglycone, abrutogenin. Other compounds of leaves are triterpenes abrusgenic acid, abruslactone A and methyl abrusgenate and flavonoids vitexin, liquirtigenin-7-mono– and diglycosides and taxifolin-3-glucoside. The roots also contain glycyrrhizin and alkaloids like abrasine and precasine besides abrine and related bases. The seeds yield alkaloids, a fixed oil, steroids, lectins, flavonoids and anthocyanins. The alkaloids of the seeds are the same as those reported from leaves. The oil content of seed is only 2.5%, which is rich in oleic and linoleic acids. β-sitosterol, stigmasterol, 5β-cholanic acid, abricin, abridin and cholesterol are the steroids present. The colour of the seeds is due to glycosides of abranin, pelargonidin, cyanidin and delphinidin. A sapogenol, abrisapogenol J, sophoradiol, its 22-O-acetate, hederagenin methyl ether, kaikasaponin III methyl ester, abrusin (8-C-glucosyl scutellarein 6,7-dimethyl ether), its 2''-O-apioside, flavones such as abrectorin and aknone (Markham et al. 1989; Choi et al. 1989) are the other constituents of seeds. Lectins are the chief constituents of seeds, the principal ones being abrins (unto 0.15% of seed). Lectins are both toxic (abrin) and non-toxic (*Abrus* agglutinin). Abrins are denoted by abrin a, b, c & d and consists of one large β-polypeptide chain (MW. 35,000) and a short α-polypeptide chain (MW. 32,000) joined by a disulphide bond. Agglutinin consists of 4 polypeptide chains; the chains are similar to those of abrins.

The roots, leaves and seeds of this plant are used medicinally. The roots and leaves are used as a substitute of liquorice in coughs and catarrhal affections. The roots possess diuretic, tonic and emetic properties and are also used in gonorrhoea, jaundice and other infections. The plant extract is one of the constituents of oral contraceptives. The seed extract is used externally in the treatment of ulcers and skin affections. Administered internally, the seeds are useful in affections of the nervous system, diarrhoea and dysentery. Abrin is extensivelly studied for its antitumour activity, where it was found to suppress tumour growth in rats and mice. Abrin is found to inhibit protein synthesis by inactivating ribosome dependant GTPase and exhibit no bone marrow suppressing affects as other cytostatic agents.

Agave americana Linn. (Agavaceae)

Kantala, American Aloe

A. americana is a short-stemmed plant bearing a rosette of long, erect, pointed, fleshy leaves with marginal spines and longitudinal white yellow streaks or bands. The plant flowers only once during its life and the flowering stem arises from the centre of the plant as a thick pole. The flowers are funnel shaped and capsules oblong clavate and beaked. A native of America, it was introduced in India for its leaf fibre.

The leaves yield saponins such as agavasaponins A-I (10 in number) based on hecogenin (aglycone), and chlorogenin, rockegenin, tigogenin, dehydrohecogenin and piscidic acid. The flowers contain chlorogenin and kaempferol and the seeds neotigogenin, hecogenin and kammagenin.

The leaves are used as laxatives and emmenagogue and in scurvy. Leaf juice is used for warts, cancerous ulcers and tumours. Roots are diuretic, diaphoretic and antisyphilitic. Hecogenin extracted from Agaves are used in the manufacture of corticosteroids β-methasone and dexamethasone.

Asparagus racemosus Willd (Liliaceae)

Shataveri, Asparagus

A. racemosus is a scandant much-branched spiny undershrub with tuberous short rootstock bearing a number of tuberous roots, stems armed with numerous recurved spines, sickle-shaped cladodes, leaves reduced to scale leaves, small white flowers in racemes and globose berries. Asparagus occurs wild and is cultivated throughout India.

The tubers contain saponins, named shatavarins 1,11,111,IV, glycoside AR-4 ($C_{45}H_{74}O_{16}$), tridecaacetyl shatavarin-I ($C_{77}H_{112}O_{36}$) and tetradeca-O-methyl shatavarin-I ($C_{65}H_{114}O_{23}$). Shatavarin-IV is based on sarasapogenin. Also present are sarasapogenin, β-sitosterol, 4,6-dihydroxy-2-O-(2′-hydroxy isobutyl) benzaldehyde and undecanyl cetonoate. Carbohydrates (53%) inclusive of mucilage, polyfructosans, free sugars and insoluble polysaccharides as well as proteins (3%) form the other components of the tubers. A polycyclic alkaloid, asparagine-A is reported from plants of Thailand. The cladodes are said to contain diosgenin and quercetin. Flowers yield a volatile oil, rutin, hyperoside and quercetin. The fruits also yield quercetin, rutin and hyperoside along with cyanidin glycosides (Joshi and Dev, 1988).

Asparagus is a tonic, demulcent and aphrodisiac and used for hyperacidity, diarrhoea and dysentery. It is a well-known galactogogue and beneficial to menstrual troubles and nervous breakdown.

Chlorophytum borivilianum Sant. & Fernandes (Liliaceae)

Sweta musali/Safed musali

C. borivilianum is a small perennial herb with 1-8 sessile, cylindrical brown to black-skinned (white after peeling) fleshy root tubers, 6-13 sessile linear radical spreading leaves, solitary terminal scape containing white flowers in clusters of 3 and a 3-angled loculicidal capsule enclosing many small black angular seeds. This is a native of western India, now widely cultivated.

The tubers contain saponins (2-3%) based on hecogenin, protein (8%) and carbohydrates (40%) mainly as mucilage.

Safed musali, the tuber, is much valued for its aphrodisiac properties. It is a valuable nervine and general tonic. The powder is useful for aphthae of mouth and throat and also in rheumatism.

Tubers of *Chlorophytum arundinaceum* Baker, *C. tuberosum* Baker, *Asparagus adscendens* Roxb. and *A. racemosus* Wild also are used as safed musali or its substitutes. *C. arundinaceum* tubers are found to contain a bibenzylxyloside, 2′,4,4′-trihydroxy-2-xylopyranosyl bibenzyl, and steroidal sapogenins such as sarasapogenin, tigogenin, neotigogenin and tokorogenin, a disubstituted tetrahydrofuran, stigmasterol and its glucoside (Tandon and Shukla,1993).

Fagonia cretica Linn. (Zygophyllaceae)

Syn. *Fagonia arabica* Linn.

Dasparsha/Duralabha (Fagonia)

A small, much-branched spiny undershrub covered with glandular hairs, 1-3 foliolate leaves (opposite) having 2 pairs of sharp stipular spines (often longer than leaves), leaflets linear sessile, the middle longest and small rose-coloured flowers (petals double the number of sepals). The fruits are glandular, pubescent pyramidal 5-partite schizocarps enclosing flat compressed ovoid seeds. Dasparsha is common in dry areas of northwestern India, Afganistan and Iran.

The aerial parts of this plant contain at least six saponins. Saponin A&B are based on nahagenin; saponin C, based on 21α, 22β-dihydroxy nahagenin; two saponins on hederagenin and the last on oleanolic acid, Also present are diterpenes fagonone, 16-O-acetyl fagonone and 7β-fagonone, chinovic acid, ceryl alcohol, β-sitosterol, fagonin, fagogenin, betulin, oleanolic acid, compesterol, β-sitosterol stigmasterol and an alkaloid harmine. Flavonoids present are ternatin, herbacetin, gossypetin, 5,7,4′- trihydroxy, 3,8,3′-trimethoxy flavone, kaempferol, 4′-OMe kaempferol and isorhamnetin. Also present are phenolic acids such as *p*-hydroxybenzoic, vanillic, syringic, *p*-coumaric and ferulic acids and alkaloids (Ansari et al. 1988; Umadevi and Daniel, 1991; Abdel-Kadar et al. 1994)

Duralabha is a blood purifier and cures fevers, asthma, diarrhoea, phantom tumour and dermatitis. It also possesses stimulant, laxative and alterative properties.

Aesculus hippocastanum Linn. (Hippocastanaceae)

Horse Chestnut

This is a tall deciduous tree with a palmately compound leaf bearing 5-7 oval to spindle-shaped leaflets, white/pink flowers borne on large terminal racemes and a spiny green fruit containing 3 rounded shiny brown seeds. This is a native of western Asia, now cultivated around the world as an ornamental.

The seeds yield saponins, flavonoids and lipids. There are over 30 pentacyclic saponins reported from this plant of which "aescin" is considered the principal saponin. But aescin is a group of saponins, based on the sapogenins proaescigenin and barringtenol, combined with angelic, tiglic or acetic acids.

Seeds are useful in varicose veins and to prevent thrombosis, thrombophlebitis, calf cramping, edema and hemorrhoids. The extract of the seeds is antiinflammatory and used as an expectorant in asthma and bronchitis, in prostrate enlargement, dysmenorrhoea and oedema. *Aesculus* is a well-known homeopathic remedy.

Trigonella foenum-graecum Linn. (Fabaceae)

Methika, Fenugreek

Fenugreek is a herbaceous annual with pinnately trifoliolate leaves having long oblanceolate dentate leaflets, white/yellow flowers singly or in pairs from axils and long (upto 15cm) cylindrical pods containing 10-20 yellow grooved trapezoid seeds. A native of Mediterranean, fenugreek is now grown for seeds and as pot herb.

The seeds contain protein (26%), fat (6%), carbohydrates (44%), flavonoids such as quercetin and luteolin (free and as glycosides) saponins based on diosgenin, tigogenin and gitogenin, trigonelline and minerals. A galactomannan is the main component of carbohydrates. Leaves yield proteins (41%) , carbohydrates (6%) and free amino acids. Also present in the leaves are saponins (similar to those of seeds), flavonoids such as kaempferol and 7,3'-diOMe quercetin (Daniel, 1989).

The seeds of fenugreek are mainly used as a galactogogue. They are carminative, a tonic and as a poultice, are useful in ulcers, and abcesses. They are used as emollient for inflammations in the intestinal tract. The mucilage is used as a tablet binder. The leaves are a pot herb, much in demand.

Glycyrrhiza glabra Linn. (Fabaceae)

Yahtimadhuh, Liquorice

This glandular hairy herbaceous perennial, 1 m in height, bears pinnate leaves having 9-17 ovate leaflets, pale blue flowers in spikes and a 3-4 seeded red-brown glabrous pod.

Liquorice, a native of southern Europe and western and central Asia, is cultivated on a mass scale in Spain, Turkey and the USSR.

The roots contain saponin glycyrrhizin (2-9%) and a mixture of potassium and calcium salts of glycyrrhizinic acid as the major constituents. Also present are some other triterpenoid saponins such as glabranin A&B, glycyrrhetol, glabrolide and isoglabrolide; flavonoids such as glucoliquiretin apioside, prenyllicoflavone A, shinflavone, shinpterocarpin and 1-methoxy phaseollin, isoflavones like formononetin, glabrone, neoliquiritin and hispaglabridin A & B; coumarins, viz., herniarin, umbelliferone and sterols, onocerin, β-amyrin and stigmasterol (Hikino, 1985; Bradley,1992; Bisset,1995).

The drug exhibits demulcent, expectorant, antimicrobial, antihepatotoxic and anti-inflammatory properties due to glycyrrhizin. It is also useful in treating peptic ulcer. Oral doses of liquorice are given for gastric, duodenal and oesophageal ulceration or inflammation, heartburn and mouth ulcers. It prevents infections of the urinary tract and is used in Chinese medicine as an emmenagogue. Glycyrrhizin extract is used for chronic hepatitis. Glycyrrhetinic acid exhibits a specific antitumour activity and is a cytostatic. It also induces phenotypic reversion, i.e. the cancer cells get converted into normal cells.

Balanites roxburghii Planch. (Simaroubaceae)

Syn. *B. aegyptiaca* Delile

Hingoli, Desert Date

B. roxburghii is a spiny evergreen tree attaining a height of 9 m with bifoliolate leaves having elliptic leaflets, small greenish flowers in axillary few-flowered clusters and ovoid woody 5-groved fruits containing an oily seed. This plant is common in the drier parts of India and is a native of Africa.

The fruit pulp contains several steroidal saponins balanitisins A-E, based on the sapogenin diosgenin, yamogenin and cryptogenin. A furostanol saponin, balanitoside along with balanitin-3 and 6-methyl diosgenin is also reported from mesocarp. The amount of total saponins comes upto 4%. The seed kernels yield balanitins 4, 5, 6 & 7 and balanitisin F. Yamogenin is the major component of certain samples of fruit epicarp. Seeds contain a fixed oil (40%), rich in oleic (50%), palmitic (24%) and linoleic (20%) acids besides protein (54%). Stem bark contains steroidal saponins deltonin, prodeltonin; furanocoumarins bergaptene and marmesin as well as two sesquiterpenes balanitol and its isomer. The root wood yields balanitisin H and stem wood, balanitisin I. Leaves also contain a saponin of diosgenin, stigmasterol and a small amount of free diosgenin (Jain, 1987; Hosny et al. 1992b).

Almost all parts of the plant are found to be anthelmintic and purgative. The fruit is useful in whooping cough, skin diseases, and for pneumonia (in the form of an application to the chest). Seeds are also used as expectorants and oxytocics.

Bacopa monnieri Wettst. (Scrophulariaceae)

Syn. *Herpestris monnieri* H.B. & K.

Brahmi, Bacopa

B. monnieri is a small creeping succulent herb, rooting at nodes with simple opposite obovate-oblong leaves, white solitary axillary flowers and ovoid capsules. Brahmi is found in marshy places throughout India.

The entire plant contains alkaloids, saponins and flavonoids. Alkaloids found are brahmine, herpestine, nicotine and other bases. The saponins located are monnierin, hersaponin, bacoside A, A_3 and bacoside B. Bacoside A is based on jujubogenin, whereas bacoside B is based on bacogenin. Betulic acid, stigmasterol and β-sitosterol are the steroids present. Luteolin and mannitol are the other compounds present (Rastogi et al. 1994).

The whole plant, known as *Brahmi*, is a very well-known brain tonic and is found to be effective in anxiety neurosis and to revitalize intellectual facility (Sharma et al. 1987). The plant is used against asthma, epilepsy, and insanity and as a potent nerve tonic, cardiotonic and diuretic. Other uses are against bronchitis, diarrhoea and rheumatism. Fifty percent of the alcoholic extract is found to the anticancerous.

Serenoa repens Small (Arecaceae)

Syn. *Sabal serrulata* Linn.

Saw Palmetto

Saw palmetto is a small prostrate palm tree, upto 4m in length, and a native of southeastern United States. Roots are located on the ventral side of the stem and are concentrated near the growing end. This palm bears palmately divided (upto 20 lobes or more) leaves (upto 1m wide) having a petiole toothed with sharp spines. The flowers are small, white, borne on the spadix and the fruit is a single-seeded bluish-black fleshy drupe.

Fruits yield volatile oil, fixed oil, steroidal saponins, tannins and polysaccharides. The fixed oil consists of 25% free fatty acids such as caproic, lauric and palmitic acids and 75% glycerides.

A concentrated extract of berries is useful in maintaining a healthy prostrate function. The extract is found to prevent testosterone from converting to dihydrotestosterone, a hormone thought to cause prostrate cells to multiply leading to an enlarged prostrate.

Taraxacum officinale Weber (Asteraceae)

Kanphul, Dandelion

T. officinale is a laticiferous perennial having a thick taproot, radical, sessile irregularly pinnatifid leaves with triangular toothed lobes, yellow flowers in ligulate heads and flattened spiny (upper half) achenes bearing white pappus hairs. Dandelion is a native of Central Asia.

The roots contain inulin, steroids such as taraxasterol, taraxacerin, taraxacin, gum and rubber.

Roots and leaves are alterative, cholagogue, diuretic, lithotropic, tonic and used for a variety of ailments connected with menopause, like mood swings. They are also used to alleviate nausea during pregnancy. In Chinese medicine, this plant is used for breast problems, including breast cancer.

Guaiacum officinale Linn. (Zygophyllaceae)

Lignum vittae, Guaiacum

G. officinale is a small evergreen tree with paripinnate compound leaves (leaflets 3 pairs, ovate trapezoidal), deep blue flowers in axillary clusters and heart-shaped capsules. This plant is a native of South America.

A number of saponins are reported from various parts. Leaves contain guaicins A to F (guaicins C, D and E based on 30-nor-Olean-12, 20/29)-dien-28-oic acid while guaicin F, based on oleanolic acid). Stem bark yields guaicins D & E, while the fruits contain guaicins F & G. Heartwood contains lignans such as furoguaiacidin and furoguaiaodin.

The wood and extracts are hepatotonic, antiinflammatory, diuretic and laxative and useful in treating gout, arthritis and other rheumatic conditions.

Catunaregam spinosa Tirveng. (Rubiaceae)

Syn. *Randia spinosa* Poir.
R.dumetorum Poir.
Xeromphis spinosa Keay.

Madana, Emetic Nut

C. spinosa is a thorny tree with dark brown bark, obtuse deciduous leaves narrowed into a short petiole, axillary thorns, solitary fragrant white flowers turning yellow and globose yellow berries enclosing flat angular seeds. This plant is common as undergrowth in sal forests.

The bark contains saponins based on randialic acid A & B, mannitol (upto 6%), 10-methyl lixoside (an iridoid) and scopoletin. Stems yield iridoids such as randinoside, galioside, deacetylasperulosidic acid methyl ester, scandoside methyl ester, geniposide and gardenoside. Roots also contain mannitol, scopoletin and a yellow dye. The leaves yield flavonoids, kaempferol, 4-OMe kaempferol and quercetin and phenolic acids *p*-hydroxy benzoic, gentisic, melilotic, vanillic and syringic acids (Thomas, 1989; Hamerski et al. 2003).

The fresh fruits contain 2-3% saponins (10% dry wt.) which are mostly concentrated in the pulp. They are dumetorins A, B, C, D, E and F (based on oleanolic acid), randioside A (on oleanolic acid), ursosaponin (on ursolic acid), two triterpenoid glycosides based on oleanolic acid and 10-methyl lixoside. Seeds contain protein (14%) and saponins based on oleanolic acid.

The bark is used in diarrhoea, dysentery and is an abortifacient, anthelminitic, antipyretic and emetic. It is also considered to be sedative and hypoglycemic. The roots are used for treating gonorrhoea and for poisoning fish. The fruit pulp, dried and powdered, is emetic and substituted for ipecac. It is also a nervine calmative, antispasmodic, anthelmintic and abortifacient.

Centella asiatica Urban. (Apiaceae)

Syn. *Hydrocotyle asiatica* Linn.

Manduk parni /Brahmi, Gotu Kola

C. asiatica is a creeping stoloniferous annual/perennial with a prostrate stem rooting at the nodes, reniform long-petioled leaves, minute reddish flowers in umbels of three and a laterally compressed fruit. It is a common weed in crop fields and moist places.

A number of triterpenoid glycosides are isolated from this plant. They are asiaticoside, indocentelloside, brahmoside, brahminoside, thankuniside and isothankuniside. These saponins are based on asiatic, indocentoic, brahmic, thankunic and isothankunic acids as aglycones. Samples from Europe are found to contain madecassocide (aglycone madecassic acid). Asiatic, indocentoic and brahmic acids along with isobrahmic and betulic acids occur in free form. *meso*-Inositol, kaempferol, quercetin, sitosterol, campesterol and stigmasterol are some other components of this drug. Also present is a volatile oil (containing β-caryophyllene, *trans* β-farnesene and germacrene - D), a fatty oil, an alkaloid hydrocotylin ($C_{22}H_{33}NO_8$) and polyacetylenes.

Gotu kola is used as **Brahmi** (*Bacopa monnieri*) at many places. This is one plant recommended for improving memory, reducing anxiety neurosis and hypertension and for treatment of leprosy and skin diseases. Externally, the paste of the fresh herb is applied for rheumatism, elephantiasis and hydrocele. It is valued as a tonic and used in asthma, catarrah, leucorrhoea, kidney problems and tuberculosis.

Costus speciosus Smith. (Zingiberaceae)

Canda, Costus

C. speciosus is a succulent herb with tuberous rhizomes, spirally twisted stem, oblong acuminate leaves pubescent below, large white flowers having reddish brown bracts in dense terminal spikes and an ovoid 3-valved capsule enclosing black seeds with a white aril. Canda can be seen in dense evergreen forests of India.

Rhizomes contain saponins dioscin and gracilin based on diosgenin (2.12%), tigogenin and β-sitosterol and a volatile oil. Roots yield aliphatic hydroxy ketones and 5α-stigmast-9 (11)-en-β-ol. The seeds also contain diosgenin.

The rhizome is bitter, astringent and used for dyspepsia, fever, cough and other respiratory diseases, diabetes, oedema, blood diseases and skin ailments and also improves the complexion. Saponins and genins cause spasmodic uterine contraction and are estrogenic. They are antiflammatory, hypotensive and bradycardic in nature.

Curculigo orchioides Gaertn. (Amaryllidaceae)

Syn. *C. malabarica* Wight.
Hypoxis orchioides Kurz.

Musali

C. orchioides is a small perennial herb having a long cylindric rootstock bearing a large number of root tubers. Leaves are basal, long, sessile linear/ lanceolate with the tips often bearing bulbils. Flowers are small yellow, borne on a short scape hidden in the leaf sheath and the fruit is a 1-4 seeded capsule. This plant is common in forests of central and western India.

Tubers contain a glycoside yuccagenin, an alkaloid lycorine, flavones and 3-MeO, 5-Ac, 31-tritriacontane.

The tubers (musali) are a well-known rejuvenative (rasayana) drug and an aphrodisiac. It is slightly bitter, viriligenic, diuretic and is useful in general debility, cough, piles, skin diseases, impotence, jaundice, urinary diseases, leucorrhoea and menorrhagia.

Tribulus terrestris Linn. (Zygophyllaceae)

Goksura/Gokhru, Tribulus

T. terrestris is a prostrate wooly perennial with paripinnate leaves (leaflets ovate/elliptic), yellow flowers, and a pentagonal spiny fruit. This medicinal plant is a common weed throughout India.

The fruits, which form the drug, contain steroidal saponins, alkaloids and steroids. The saponins, terrestrosins A-E, desgalactotigonin, F-gitonin, desglucolanatigonin and gitonin are based on sapogenins diosgenin, hecogenin and neotigogenin. β-Sitosterol, stigmasterol, terrestiamide (a cinnamic acid amide) and 7-methyl hydroindanone are the other compounds reported. Leaves contain kaempferol, isorhamnetin, vanillic acid, syringic acid, melilotic acid and *p*-coumaric acid. (Ren et al. 1995; Yan et al. 1996).

Gokhru is a well known diuretic and antiurolithiatic. It is also nephroprotective and a cardiac stimulant.

Actaea racemosa Linn. (Ranunculaceae)

Syn. *Cimigifuga racemosa* Nutt.

Black Snakeroot, Black Cohosh

Black cohosh is an erect herb with a stout blackish hard and knotty rhizome bearing many stout ascending branches, tripinnate leaves (leaflets ovate serrate/dentate), cream-coloured flowers on tall racemes and an etaerio of follicles. *A. racemosa* is a native of N. America.

Roots and rhizomes contain triterpene glycosides actein, acetylactal, 27-deoxyactein, cimigenol and cimifugoside, quinazoline alkaloids, isoflavone (formonetin,) tannins and gallic, isoferulic and salicylic acids.

Roots and rhizomes are widely used in many gynecologic disorders, especially menopausal and menstrual dysfunction. The terpene glycosidal fraction is found effective in gonadotropin release in menopausal women and thus reduces premenstrual discomfort, dysmenorrhoea or other menopausal ailments.

3.6d Cardiac Glycosides

Cardiac glycosides also are triterpenoid glycosides but differ from saponins in having an unsaturated lactone ring at C-17 of the nucleus, a *cis* juncture of the rings C & D, an additional hydroxyl group at C-14 and the unique sugars they possess. They are also soluble in water and are surface-active agents, with soap-like properties. Cardenolides, being C_{23} compounds, possess a 5-membered lactone ring (e.g. sarmentocymarin) and bufadienolides (scilladienolides) are C_{24} compounds with a doubly saturated 6-membered lactone ring (e.g. Scilliroside). Both cardenolides and bufadienolides are mutually exclusive.

As the name suggests, cardiac glycosides elicit specific and powerful action on the cardiac muscle, the prime reason why natives used them as arrow poisons. This activity increases with the increase in hydroxyl groups. Their genins also exhibit these properties.

All the sugars, linked at C_3 position, possess methyl and/or methoxy groups (e.g. digitalose) and the number of sugar units present in a glycoside ranges from 1 to 4.

Digitalis purpurea Linn. (Scrophulariaceae)

Digitalis

A perennial, but biennial in cultivation, *Digitalis* reaches a height of 1 m and bears both long-petioled, radical leaves and sessile stem leaves. The large pendulous purple flowers are borne on a one-sided raceme. The bilocular capsule encloses numerous seeds. This plant, a native of southern and central Europe, is now extensively cultivated in all temperate parts of the world.

The rapidly dried leaves, which form the drug *Digitalis*, contain more than 40 cardiac glycosides based on 4 genins, digitoxigenin, gitoxigenin, gitaloxigenin, and gitaligenin along with their formyl or acetyl derivatives. The glycosides may be: (1) monoglycosides like verodoxin (gitaloxigenin digitaloside) and odoroside H (digitoxigenin digitaloside); (2) diglycosides like gitalin (gitaligenin + 2 x D-digitoxose) and digitalinum verum (gitoxigenin + digitalose + glucose); (3) triglycosides like gitoxin (gitoxigenin + 3 x digitoxose) and gitaloxin (16-formyl gitoxin; or (4) primary or tetraglycosides like purpurea glycoside A. (digitoxigenin + 3 x digitoxose + glucose).

The two primary glycosides—purpurea glycosides A & B—form the chief constituents of fresh leaves. But in dried leaves, triglycosides such as digitoxin and gitoxin are predominant. The triglycosides are probably derived from primary glycosides by an enzymatic splitting taking place during the process of drying. The glycosides of digitoxigenin and gitaloxigenin possess more activity than those of gitoxigenin. It is often observed that in the leaves, gitaloxigenin readily gets hydrolyzed to produce gitoxigenin and formic acid.

In addition to these compounds, leaves contain a group of glycosides, numbering a dozen, which resemble the cardenolides but for the lactone ring, named digitenolides. The known aglycones are diginigenin, digacetigenin, purprigenin and purprogenin and they occur as monoglycosides (diginin = diginigenin D-diginoside) or tri-glycosides (purpronin = purprogenin-digitoxosides). These compounds do not possess the cardiac activity.

Also present in the leaves are saponins, gitoxin- (gitogenin + 4 x galactose + xylose) and digitonin (digitogenin + 2 x glucose + 2 x galactose + xylose) exhibiting some of the properties of cardenolides and a few anthraquinone derivatives like 3-methyl alizarin and 3-methoxy, 2-methyl anthraquinone.

Digitoxin is a cardiotonic increasing the tone of the cardiac muscle. Gitalin either alone or with other glycosides, are often used in treatment of congestive heart failure. All these glycosides improve the rhythm of heartbeats, making the contraction of the heart more powerful and help the heart to pump the blood at the time of cardiac failure.

Along with some of the active principles of leaf, *Digitalis* seeds contain some other cardenolides like digifucocellobiosides (digitoxigenin + fucose + glucose), gitosin and neogitosin; and saponins like gitonin and digitonin. Seeds are used as a substitute for the leaf occasionally.

Digitalis lanata Ehrh. yields more than 70 glycosides based on the four genins reported from *D. purpurea* and two new ones digoxigenin and diginatigenin. Digoxin, a triglycoside and the primary glycosides, lanatosides A, B, C, D and E are major ones. It is observed that in all the glycosides showing cardiac activity, the first three sugar residues are always digitoxose. Though some genins are similar to those of *D. purpurea,* their primary and triglycosides differ in having an acetyl group attached to the third digitoxose residue. Lanatosides and digoxin are valuable cardiotonics.

Convallaria majalis Linn. (Liliaceae)

Convallaria

A native of Europe, E. Asia and N. America, this low-scapose rhizomatous plant produces only 2 oblong-oval leaves (15-20 cm long) situated one above the other. The flowers are white, borne on racemes. The fruit is a red berry containing a few seeds. The aerial parts, collected before the opening of the flowers, and the rhizome with roots form the commercial drug.

Of the 25 or more glycosides isolated from the drug, the important ones are convallatoxin (strophanthidin + rhamnose), convalloside (convallatoxin + glucose) and convallatoxol (strophanthid-19-ol + rhamnose). Also present in the drug are various flavones. Roots contain a saponin-convallamaroside also. The flowers are the source of 'Lily-of-the-valley' flower oil.

The drug is a cardiotonic but has lesser action than **Digitalis**. Convallatoxin is an effective diuretic. Flower glycosides are found to strengthen and regulate heart action and in dropsy, they assist urine secretion.

Strophanthus kombe Oliv., *S. hispidus* DC. (Apocynaceae)

Strophanthus

The seeds of these tall shrubs, native to Africa, provide the drug Strophanthus.

The dry seeds yield about 8-10% strophanthin or K-strophanthin, a mixture of more than 10 glycosides. The chief constituent is

K-strophanthoside (strophoside), a triglycoside consisting of the genin, strophanthidin, cymarose and 2 molecules of glucose. Two other major glycosides are cymarin and K-strophanthoside B. The genins present are strophanthidin, alloperiplogenin, strophanthid-19-ol, strophanthidic acid and periplogenin. The other constituents of the seed are about 30% fixed oil, nitrogenous bases like trigonelline and choline, resin and mucilage.

Strophanthus is the cardiotonic drug preferred to *Digitalis* in Europe.

The seeds of *Strophanthus gratus* Oliv., a native of Africa, provide another cardiotonic drug Ouabain (g-strophanthin), a rhamnoside of ouabagenin. One of the most rapidly acting cardenolide, ouabain is also obtained from the wood of *Acokanthera* sp.

Asclepias curassavica Linn. (Asclepiadaceae)

Kakatundi, Blood Flower

This is an erect much-branched perennial with a woody rootstock, lanceolate leaves, orange-red flowers in cymes and a pair of follicles containing comose seeds. It is a native of tropical America, now cultivated as an ornamental plant.

The leaves of Brazilian plants yield cardenolides such as clepogenin, curassavogenin, ascurogenin, carotoxigenin, uzarigenin, coroglaucigenin and a glycoside uzarin. But the leaves from Indian plants are devoid of the first four compounds but are found to contain 22 cardenolides of which calactin, calotropin, calotropagenin and asclepin are the major compounds. The Mexican samples also yielded a cytotoxic compound, calotropain. The roots are found to contain vincentoxin (asclepiadin).

All parts of the plant are used in medicine. Root is considered emetic, cathartic, astringent and a remedy for gonorrhoea and piles. Roots are used as substitute of ipecac. The whole plant is used as emetic, styptic and purgative and the extract is found to inhibit carcinomatous cells of human nasopharynx. The plant, in the form of powder, balm or enema, is used to destroy abdominal tumours. The latex is used against warts and corns.

Carissa carandas Linn. (Apocynaceae)

Karaunda, Christ's Thorn

C. carandas is a small laticiferous tree with dichotomous branches armed with a simple or forked pair of thorns at the origin of branches. The leaves are elliptic and flowers are white/pink, faintly scented in terminal corymbose cymes. The fruits are pink ellipsoid berries containing eight seeds. A native of south Asia, this plant is grown for edible fruits.

The roots contain cardiac glycosides based on odoroside H and terpenoids such as carissone, carindone, carinol and related compounds. The leaves contain lupeol, β-sitosterol and ursolic acid. Flowers yield a volatile oil rich in phenyl ethyl acetate (60%), benzyl acetate (22%) and linalool (8%). The fruits are rich sources of Vit.C and yield carissol (an epimer of β-amyrin), lupeol, β-sitosterol and organic acids like oxalic, tartaric and citric acids. The seeds yield a fixed oil rich in palmitic (66%), arachidic (21%) and stearic (10%) acids.

All parts of the plant are attributed with medicinal properties. Roots are cardiotonic, anthelmintic and effect a prolonged blood pressure lowering effect. The leaves are recommended in intermittent fevers. Ripe fruits are antiscorbutic, useful in bilious complaints and revealed cytotoxic activity in their fat-soluble fraction.

Cerbera manghas Linn. (Apocynaceae)

Syn. *C. odollum* Gaertn

Odollum, Cerbera

C. manghas is a small laticiferous tree with coriacious linear-lanceolate leaves, white yellow/red throated fragrant flowers in terminal cymes and subglobose green drupes containing a single seed. A native of tropical Asia, this is found in coastal areas.

All parts of the plant contain cardiac glycosides. The stem contains neriifolin, seven glycosides based on digitoxigenin, two glycosides based on tanghmigenin and one based on oleagenin and free tanghinigenin. Also present are lignans, (-)- olivil, (+)-cyclo olivil, 5,5′′′-*bis*-olivil and similar dimers. Iridoids such as 10- and 11- carboxy carbinal also are isolated from the stem bark. The leaves are found to contain cerleaside A and B as the major cardenolides, and 5 glycosides of digitoxigenin, 5 of tanghinigenin, neriifolin, thevetin B, thevoside and thevirioside as minor components. Lignans present here are glucosides of olivil. Also present are quercetin, rutin, clitorin, nicotiflorin, bornesitol, and three normonoterpenoids, cerberidol, epoxy cerberidol, cyclocerberidol and their four β-allopyranosides and two iridoids, 10-O-benzyl thevoside and 10-dehydrogeniposide. Seeds yield cerleaside A, 2′-O-acetyl cerleaside A, 17α-neriifolin, 17β-neriifolin, cerberin, neriifolin, thevoside, thevirioside, odollin, odollotoxin etc besides a pale yellow oil having palmitic (32%), oleic (38%), linoleic (18%) and stearic (11%) acids (Abe et al., 1988, 1989; Laphookhieo et al., 2004).

The bark and leaves are cathartic and emetic. Seeds are fish poisons. Seed oil is used as hair oil and as a rubefacient to cure cold.

Calotropis gigantea Ait.f. (Asclepiadaceae)

Arkah, Milkweed

C. gigantea is an erect tomentose laticiferous shrub reaching a height of 2 m with large ovate subsessile leaves, white/purple-tainted flowers in umbellate cymes and a pair of subglobose inflated follicles enclosing broadly ovate comose seeds. This is a very common weed occurring in the dry regions of India.

The root bark, which constitutes the drug, yields cardiac glycosides such as gigantin, giganteol, isogiganteol, calotroposides A-G, steroids like α- and β-amyrins, taraxasterol and its derivatives, β-sitosterol and a wax. The latex, found in all parts of the plant, contains caoutchouc (rubber), resin, cardiac glycosides gigantin, calotropin, uscharin, calotoxin, calactin and uscharin (all based on the genin calotropagenin), proteases like calotropain, calotropin D_1 and D_2 , calotropain F_1 and F_2 and enzymes like invertase. Flowers yield a resin containing resinols, α- and β-calotropeol, giganteol, calotropin, β-amyrin, flavonoids and fatty acids. Seeds contain a fatty oil (30%; rich in oleic acid and linoleic acid), stigmasterol, melissyl alcohol and laurane. Leaves contain free sapogenins, β-amyrin, β-sitosterol, taraxasterol, ψ-taraxasterol, tannins and resin. The flowering tops contain holarrhetine and cyanidin glycoside.

Root bark is a substitute for ipecac and in small doses, it acts as a diaphoretic and expectorant. It is used in cures for leprosy and eczema. A 50% ethanolic extract is found to be anticancerous and low doses of crude methanolic extract act as a hypertensive and cardiotonic. The latex is a drastic purgative and emetic and induces abortion. The flowers are useful in cough, cold, catarrah and asthma. The fibre is used for ropes, clothes and as a filling fibre. The whole plant is used as an insecticide (against white ants and larvae of *Culex* and *Anopheles*), fish poison and possesses both antifungal and antibacterial properties.

In Indian systems of medicine, the tender fresh leaves are used to cure migraine and fits and convulsions in children.

Calotropis procera Ait.f.

Alkarka, Dead Sea Apple

C. procera is similar to *C. gigantea in* overall appearance but differs in having a broad, but short corona which is equal to or longer than the staminal column and not having the two auricles below the apex of corona. The height of the entire staminal column is about 0.5 to 0.8, i.e. less than a centimeter and the stigma is almost sunken in the corona lobes. [In C. *gigantea*, the corona is narrower and longer, much curved back to the staminal column towards halfway down, more than 1cm in height, but always lower than the gynostemium (stigma always elevated up) and possesses two obtuse structures (auricles) at the apex of each lobe.]

The root bark, which constitutes the drug, contains α- and β-amyrins, taraxasterol and its isomers, isovalerates and acetates, giganteol, β-sitosterol and a wax. The latex, present all over the plant, consists of caoutchouc, resin and cardiac glycosides, proteases and steroids. The cardiac glycosides are calotropin, uscharin, calotoxin, calactin, uscharidin (all based on the genin calotropagenin), voruscharin, proceroside and two genins, uzarigenin and syriogenin. The proteases encountered are calotropain F_1, F_2, calotropin D_1 and D_2. α-Amyrin, β-amyrin and β-sitosterol are the steroids present. Flowers contain α-, β-amyrins, stigmasterol, β-sitosterol, multiflorenol, cyclosadol, procerasterol and calotropenyl acetate (Khan & Malik, 1989).

C. procera is found to possess almost all medicinal properties of *C. gigantea* and used likewise.

Corchorus olitorius Linn. (Tiliaceae)

Jew's mallow

C. olitorius, a source of jute fibre, is a much-branched annual of 1m, with elliptic lanceolate serrate (lower serratures on both the sides prolonged to a filiform appendage) leaves, small pale yellow flowers and cylindric, 10-ribbed beaked capsules containing many small trigonous seeds. A common weed, this plant is often cultivated for the fibre from its stem.

Seeds contain cardenolides such as olitoriside A, erysimoside, coroloside, helveticoside, corchoroside A, evonoside, strophanthidol and strophanthidin. Roots yield a triterpene, corosin and β-sitosterol. Leaves contain kaempferol, quercetin, p-hydroxy benzoic acid, vanillic acid, syringic acid and melilotic acid.

Seeds are ecbolic in nature and are used as a cardiotonic. The cardenolides are very effective in acute and chronic cardiac insufficiency, peroxystic tachycardia and tachyarythmia.

Pergularia daemia Choiv. (Asclepiadaceae)

Syn. *Daemia extensa* R.Br.
Pergularia extensa N.E.Br.

Kurutakah, Pergularia

P. daemia is a perennial twiner with a milky latex, stem clothed by spreading hairs, broadly ovate (base cordate) pubescent leaves in pairs, small greenish white flowers (with gynostemium and double corona) in lateral cymes, and a pair of echinate lanceolate follicles containing ovate seeds velvety pubescent on both sides. This common weed is found all over India, S.E.Asia and Africa.

The stem and seeds of Pergularia contain cardenolides such as calactin, calotropin, calotropagenin (from seeds) uzarigenin and coroglaucigenin (from stems). Leaves contain kaempferol, vanillic acid, syringic acid, *p*-hydroxybenzoic acid, gentisic acid, ferulic acid, sinapic acid and *o*-coumaric acid (Daniel & Sabnis, 1982).

The plant is used for vesical calculus, dysurea and anurea. Leaf juice is emetic and expectorant and used in asthma, rheumatism, menstrual disorders and diarrhoea. Leaves are also used externally for rheumatic swellings.

Nerium oleander Linn. (Apocynaceae)

Syn. *Nerium indicum* Miller,
N. odorum Soland.

Karavirah, Oleander

This is an evergreen laticiferous shrub with linear-lanceolate coriaceous leaves in whorls of three, large showy scented flowers (colours vary from red/ yellow/white; with corolline corona, twisted petals and sagittate anthers) in terminal cymes and a pair of cylindrical follicles containing linear, ribbed comose seeds. A native of western Himalayas, oleander is cultivated in Asia, Europe and Africa.

The bark yields a number of cardiac glycosides designated as odoroside A, B, D, F, G, H, K-M and odorobioside K based on digitoxigenin (odoroside A, D, F, G & H), uzarigenin (odoroside B, K and odorobioside K) and 16-anhydrodigitoxigenin (odoroside L & M). Also present are scopoletin, scopolin, tannins, fatty oil and wax. Leaves contain oleandrin (a cardenolide based on acetyl gitoxigenin), nerium E (16-deacetyl anhydro-oleandrin), nerium F (16-anhydrodigitoxigenin), ursolic acid, oleanolic acid, neriodin, rutin, vanillic acid, syringic and salicylic acid.

The root, the official drug, is considered to be highly toxic, but in controlled doses, it is a cardiotonic. It is anthelmintic, diaphoretic, carminative and used for ulcers, leprosy and skin diseases. Leaves and roots, as an external paste, is useful in leprosy and snake bites. Leaf juice is useful in ophthalmia and improving eyesight.

Leptadenia reticulata W & A. (Asclepiadaceae)

Jivanti, Leptadenia

Jivanti is a laticiferous shrubby twiner with small elliptic-oblong leaves, small greenish yellow flowers in axillary many-flowered globose cymes (flowers with a gynostemium and pollinia) and a pair of subwoody cylindric, shortly curved beaked follicles containing ovate-oblong comose seeds. Jivanti is common in Western Ghats and N.E.Asia.

Stems and roots yield wax consisting of long-chain alcohols (C_{28}-C_{34}) and acids (C_{28}-C_{34}), sterols such as stigmasterol, γ-sitosterol and a fructosan of 7-8 units. Leaves yield apigenin, luteolin, vitexin, isovitexin and phenolic acids like vanillic, syringic, *p*-hydroxybenzoic, gentisic, ferulic, *p*-coumaric and *o*-coumaric acids (Daniel and Sabnis, 1982).

Roots are considered a rasayana (tonic) drug, useful for vitalizing the body. *L. reticulata* is an aphrodisiac, rejuvenative, galactogogue and cures weakness, cough, dysentery, night-blindness and tuberculosis. Leaves are useful in treating skin diseases.

Thevetia peruviana Merril (Apocynaceae)

Syn. *T. neriifolia* Juss.

Ashvaghna, Yellow Oleander

This is a large evergreen laticiferous shrub reaching a height of 6m, with linear acute leaves (10-15cm long), bright yellow flowers in terminal cymes and fleshy triangular fruits containing 4 seeds. A native of tropical America, it is now widely cultivated everywhere for the foliage and flowers.

All parts of the plant contain cardiac glycosides; seeds being the richest source of these compounds (containing upto 7 times) than the other parts. The various compounds isolated are cereberoside (thevetin B), 2′-O-acetyl cerberoside and neriifolin, 2′-O-neriifolin (cerberin) all based on the genin, digitoxigenin; thevetin A (19-oxocerberoside) and peruvoside (19-oxonerifolin), both based on cannogenin; thevenerrin (19-oxyneriifolin-based on cannogenol) and peruvosidic acid (based on cannogenic acid.). All of them are triosides or monosides. Also present in the seed are a fatty oil (67%) rich in oleic (64%) and palmitic (17%) acids. Bark is found to contain neriifolin and peruvoside along with lupeol acetate. Roots yield thevetin and neriifolin. Leaves contain α- and β-amyrins and flavonoids. Latex contain caoutchouc (rubber) and a resin.

All the glycosides are used as cardiotonics, the most preferred ones being peruvoside and thevetin. Bark is a bitter cathartic and emetic. Leaves are also purgative and emetic. Seeds are used as an abortifacient and as a purgative in rheumatism and dropsy.

Helleborus niger Linn. (Ranunculaceae)

Black Hellebore

This perennial herb, with unbranched underground stem and roots of thick fibres, is a native of central Europe. The scape-like flowering stem is 15-20 cm long with small deeply divided leaves and pale purple solitary flowers. The papery follicles are many seeded.

Rhizomes and parts of the aerial stem constitute the drug. Of the few glycosides it contains, hellebrin (hellebrigenin + rhamnose + glucose) shows the maximum cardiotonic activity:

Hellebrin is supposed to have approximately twenty times powerful action compared to **Digitalis**, and so is extensively used in veterinary practices.

Drimia indica Jessop (Liliaceae)

Syn. *Urginea indica* Kunth.

Jungly Piaz, Indian Squill

This is a herb with tunicate bulbs (5-10cm in diam.), flat, radical, linear sub-bifarious leaves, brown flowers (appearing before the leaves) in a raceme, borne on an erect brittle scape (upto 50cm long) and ellipsoid, trigonous capsules containing many flattened black elliptic seeds. *D. indica* is common in Western Ghats, Himalayas and tropical Africa.

Bulbs contain bufadienolides (maximum in dormant stage of the bulb) such as proscillaridin A, scillopheoside and anhydroscilliphaosidin, sterols like sitosterol and mucilage (50%, consisting of mannose, glucose and xylose).

This is used as an expectorant, cardiac stimulant and diuretic. It is also useful as an anticancer and hypoglycaemic drug and for skin diseases. This is used as a substitute to *Digitalis* when the patients show hypersensitivity to the former drug.

3.7 Carotenoids

Carotenoids, the C_{40} tetraterpenoid polyenes, are formed by tail-to-tail condensation of two diterpene molecules. They are extremely widely distributed, present in all plants and in a few animals through dietary intake. The hydrocarbon pigments are known as carotenes and their oxygenated derivatives, xanthophylls. All the carotenoids are coloured yellow or red, though a few, like phytoene and phytofluene, possessing lesser double bonds, are colourless. In most of the carotenoids, double bonds are all *trans*. Higher ring systems are absent in carotenoids and all of them are either acyclic monocylic or bicyclic. Lycopene is the acyclic basic molecule from which all the other carotenoids are derived. Cyclization of lycopene at one end gives γ-carotene while cyclisation at both ends produces bicyclic β-carotene. Xanthophylls may contain substituents like alcohol (monohydroxy-lutein; dihydroxy- zeaxanthin), epoxy (lutein) or, in rare cases, keto (rhodoxanthin) or carboxyl (crocetin). A phenolic carotenoid, 3,3′-dihydroxy isorenieratene, is reported from *Streptomyces* sp. Generally cyclization in carotenoids produces a cyclohexane ring, but capsanthin is peculiar in having a cyclopentane ring at one end. Most of the xanthophylls occur as esters of fatty acids (palmitic, oleic or linoleic acids). Glycosides of carotenoids with rhamnose or glucose are reported from various algae (Herzberg and Liaan-Jensen, 1971). In higher plants, crocin is the only glycoside known.

β-carotene is an essential dietary requirement as provitamin A. One molecule of β-carotene is found to give rise to only one molecule of C_{20} retinol (Vit. A). α-, and γ-carotenes as well as cryptoxanthin containing a β-ionone residue also can give rise to vitamin A. All the carotenoids are excellent antioxidants. Lycopene is known for its role in treating prostrate cancer. Crocin and bixin are other two carotenoids (or derived carotenoids due to the lesser number of carbon atoms) having marked medicinal properties.

Lycopersicon esculentum Mill. (Solanaceae)

Rakthamachi, Tomato

This is a pubescent spreading annual with pubescent unevenly pinnate curled leaves, yellow flowers in extra axillary cymes, large orange/red globose berries containing many flat kidney-shaped seeds on a fleshy placentum. It is a native of Peru-Ecuador, now widely cultivated throughout the world.

The fruits contain protein (2%), carbohydrates (4.5%) minerals, pectin, citric, oxalic and malic acids, carotenoids and glucoalkaloids tomatine and solanine (in traces). The carotenoids present include lycopene, β-carotene, lutein and related compounds.

Tomato is a carminative, blood purifier and general tonic. Lycopene is highly recommended for prostrate cancer and an antioxidant of great demand.

Carica papaya Linn. (Caricaceae)

Papeeta, Papaya

Papaya is a quick-growing, tall, soft-stemmed laticiferous tree having a hollow trunk bearing leaf scars, a crown of large long-petioled, deeply palmatifid leaves and unisexual dioecious flowers. The male flowers are small yellowish-white, borne in many-flowered densely pubescent cymes, while the female flowers are large, solitary or in few-flowered racemes. The fruit is a large berry mostly obovate with a hollow centre and enclosing many black reticulate seeds enveloped by a transparent aril. Papaya is a native of southern Mexico, now cultivated everywhere.

The ripe fruits are a good source of pectin, carotenoids like β-carotene, cryptoflavin and violaxanthin. Other carotenoids present are phytoene, γ-carotene, chrysanthemaxanthin and neoxanthin. Vitamin C, sugars, citric, malic and butanoic acids, volatile oil consisting of methyl butanoate, linalool, benzyl isothiocyanate and benzyl glucosinolate, alkaloid like carpaine are the other constituents. Raw papaya is a good source of pectin and proteolytic enzymes papain and chymopapain. The latex, generally collected from fruits, is the main commercial source of papain. Besides papain and chymopapain A, B & C, the latex contains sulfhydryl compounds, glutamine cyclotransferase, peptidase and lysozyme. Seeds yield protein (8%), fatty oil (9.5%, rich in oleic acid), phospholipids, carpaine, and glucosinolates like caricin, benzylglucosinolate, and glucotropaeolin and benzyl isothiocyanate. Roots contain carposide and myrosin and the leaves yield carpaine, pseudocarpaine, dehydrocarpezine, choline, vitamin C and vitamin E.

All parts of the plant are of medicinal value. The unripe fruit is a laxative and diuretic, while the ripe fruit is stomachic, galactogogue, diuretic and is effective in dysentery, chronic diarrhoea, bleeding piles, dyspepsia and as an emmenagogue. Seeds are emmanagogue, carminative and vermifuge. Carpaine shows antitumour and antitubercular activity.

Crocus sativus Linn. (Iridaceae)

Kesar, Saffron

The dried stigmas and tops of styles of this perennial herb constitute the saffron of commerce. The plant has small globose corms bearing numerous, radical, linear, ciliate leaves. The lilac/purple flowers are borne in 2-valved spathes. The styles of flowers are branched and blood red in colour.

The colour of saffron is due to a diterpene croein, which is the gentiobiose ester of crocetin—a diterpene acid having a structure resembling carotenoids. The bitter taste of saffron is attributed to picrocrocin, a glycoside yielding glucose and safranal on hydrolysis. Safranal is dihydro β-cyclocitral and is the main odoriferous constituent. The bulbs contain four isolectins. Tepals yield myricetin, quercetin, kaempferol, delphinidin, petunidin, astragalin and helichrysoside. Pollen is found to contain crosatosides A and B and kaempferol glycoside.

Saffron is a colouring and flavouring agent. Saffron extract shows cytotoxic and antimutagenic activities and anti-tumour activities against ascites tumours in mice. It is also used for its mild hypotensive properties, as an emmenagogue, in amenorrhoea, dismenorrhagia, hysteria and to prevent premature ejaculation. Crocin is used to stabilize light-sensitive drugs.

Nyctanthes arbor-tristis Linn. (Oleaceae)

Harsinghar, Nyctanthes

The dried corolla tube of this plant, native to India, is used to adulterate saffron. The pigment here too is crocin. The flowers contain a volatile oil resembling peppermint.

Bixa orellana Linn. (Bixaceae)

Shonapushpe, Annato

Annato is a small tree with cordate acuminate long-petioled leaves, moderately large pink/white flowers in terminal panicles and red/green prickly capsules containing many trigonous seeds having a red pulpy testa. A native of tropical America, Annato is commonly found cultivated.

The red pulpy seed coat contains a group of pigments, with bixin, a sesterpene (having a carotenoid skeleton) amounting to 70-80%. Other pigments present are orellin, methyl bixin, β-carotene, cryptoxanthin, lutein and zeaxanthin. Isobixin is the *trans*-form of the unstable naturally occurring *cis*-bixin. The seeds yield protein (17%), carbohydrates (14%), mucilage (4.5%), a fixed oil, an essential oil, oleoresin, phytosterols and alkaloids. The oleoresin which amounts to 1% of the seed contain all-E-geranyl geraniol (57%), farnasyl acetone, geranyl octadecanoate and volatile oil contains ishwarone (30%), α-pinene, β-pinene, etc. The dried leaves yield a gum and an essential oil (similar to that of seeds) consisting of a sesquiterpene, bixaghanene, ellagic acid, flavones such as apigenin, luteolin along with their bisulphates and glucosides, isoscutellarien, bixorellin and a steroidal sapogenin. The roots contain a triterpene acid, tomentosic acid (Jondiko & Pattenden, 1989).

Annato dye is a widely used food colour. The pulp surrounding the seed is used as a haemostatic, antidysenteric, diuretic and febrifuge. It is also recommended for epilepsy and skin diseases. The seed oil is used for treating leprosy. The leaves are used for bronchial infections (as a gargle), jaundice, dysentery, tumours and as a febrifuge. The gum from leaves is used for gonorrhea and liver complaints. Roots are antiperiodic and antispasmodic.

4

Phenolics

The term 'phenolics' circumscribes all the compounds which possess an aromatic ring bearing a hydroxyl group or its substituents. This group, considered insignificant and rare, attracted attention after its discovery in lignin and the various physiological responses it stimulates in plants and animals. Phenols are now found to have a universal occurrence, though sometimes in minute quantities. The several hundreds of phenols discovered are classified mainly on the number of carbon atoms they contain and the biosynthetic pathways by which each group is derived. Most of the phenols are derived from 5-dehydroquinic acid. But at times, the incorporation of acetate units (anthraquinone), isoprenoid chains (gossypol, ubiquinone etc), or amino groups results in the formation of highly complex phenols. The exact roles of these compounds in plants are not known with certainty but it is assumed that every compound discussed here has a definite role in the structure and/or function of the plant. In the same way, we do not know much about the structure of many polyphenols like lignins, melanins and tannins. As the phenols are considered harmful(?) if brought in direct contact with the protoplasm, the phenolic function is often blocked by glycosylation, methylation or esterification. The glycosylation makes them water soluble and helps in cellular transport.

4.1 Simple Phenols—C_6/C_7

These compounds, containing a single aromatic ring, include all the phenolic alcohols, aldehydes, ketones and their glycosides. They are colourless solids easily oxidized on exposure to air or in slightly alkaline conditions. Free phenols tend to occur in woody tissues, while glycosylated phenols are common in metabolically active sites (Hopkinson, 1969).

About a dozen phenolic alcohols are known, occuring as glycosides. Phenol itself is reported from the essential oils of pines, tobacco and *Ruta*. Catechol is present in the leaves of *Ephedra*, scales of onion and rind of lemon. Resorcinol is isolated from onion scales and tea leaves, while phloroglucinol (as a glucoside) occurs in *Citrus* fruits, and as a tautomeric triketone in many members of Pteridaceae and in the essential oils of certain angiosperms.

Hydroquinone is the most widely distributed compound in this group, reported from many members of the Rosaceae, Proteaceae and Asteraceae.

Salicin, the phenolic β-D-glucoside of salicyl alcohol, is a common constituent of higher plants. Populin and glycosmin—two derivatives of salicin—are reported from *Populus* and *Glycosmis,* respectively. The phloroglucinol derivatives of *Aspidium* (Male fern) are known for their anthelmintic properties, while that of hops (*Humulus lupulus* L.) with isoprenoid side chains are the flavoring and preservative (?) substances in beer. Urushiol, a catechol derivative with a C_{15} side chain, is the vesicant principle of *Rhus toxicodendron* L. Glucovanillin, which yields vanillin (vanillic aldehyde) the odoriferous principle of vanilla, is obtained only from monocots. Piperonal is another odoriferous principle occurring in heliotrope. β-D-Glucosides of gentisic alcohol and *p*-hydroxy acetophenone are present in *Salix.*

4.2 Phenolic Acids — C_7, C_8, C_9

This group includes both the benzoic and cinnamic acids (cinnamic acids are described in phenyl propanes). The common benzoic acids, *p*-hydroxy benzoic, vanillic and syringic acid residues are present in lignin and so located in all angiosperms. Gymnosperm lignin lacks syringyl residues. Gentisic and protocatechuic acids are two other phenolic acids common in higher plants. Salicylic and *o*-pyrocatechuic acids are frequently located in Ericaceae. Gallic and digallic acids, along with the dimeric ellagic acid, form the non-sugar (aglycone) component of gallo-and ellagitannins.

A few phenyl acetic acids also are reported from plants. 2-Hydroxy and 4-hydroxy phenyl acetic acids are reported from *Astilbe* and *Taraxacum,* respectively, while 3, 4-dihydroxy and 2, 5-dihydroxy acids are mould metabolites.

4.3 Acetophenones

These are C_6-C_2 ketones (aromatic ketones). A few of them are found to be widely distributed. Of the 15 known compounds, acetophenone itself forms the most widely distributed one. Some of these ketones, like acetovanillone, paeonol and 4-hydroxy acetophenone, occur as glycosides.

4.4 Phenyl Propanes—C_6-C_3

These compounds occur both in open and closed chain forms with or without combining with sugars. Some of the open chain forms are most widespread among the phenolics. *para*-Coumaryl alcohol, coniferyl alcohol and sinapyl alcohol (resembling *p*-hydroxy benzoic, vanillic and syringic ring structures) are well-known components of lignin and are present in free form in cambial saps of the higher plants. Phenyl propenes, a group of volatile components, occur associated with essential oils of a number of spices. Of the various cinnamic acids known, caffeic, *p*-coumaric, ferulic and sinapic acids are abundant in the leaves of higher plants free or in combination with anthocyanins. Depsides are esters formed of two or more phenolic acids. Depsidones are depsides with one O-linkage making a third ring. These compounds are extremely frequent in fungi while their abundance in higher plants is yet to be proved. Chlorogenic acid (3-O-caffeöyl-quinic acid), by far, is the most common representative of this group. Chicoric acid of chicory contains two molecules of caffeic acid and one mol. of tartaric acid. Chicoric, chlorogenic and caffeic acids belong to a new class of HIV medications known as **integrase inhibitors**, which block the enzymes that allow viruses to invade cells and "integrate" their DNA with the cell's DNA.

4.5 Benzophenones

Benzophenones are C_6-C_1-C_6 compounds with a very limited distribution among higher plants. Only eight such compounds are known till date. Cotoin (2, 6-dihydroxy-4-methoxy-benzophenone) and hydrocotoin (6-hydroxy-2, 4-dimethoxy benzophenone), found in coto bark (*Aniba*), are used as astringents in medicines. *Morus* wood contains a yellow dye maclurin (2, 4, 6, 3', 4'-pentahydroxy benzophenone). Some of the hydroxy benzophenone derivatives are condensed tannins.

Dryopteris filix-mas Schott (Polypodiaceae)

Aspidium/Male Fern

The rhizome, frond bases and apical buds of male fern, indigenous to Europe, Asia and America, constitute the drug Aspidium. The brownish black rhizome, 20-50 cm long and often as thick as an arm, is covered by the persistent leaf bases. The long bipinnate leaves, arranged spirally around the stem, are 1-2 m long with round leaflets and brown sori crowded on the lower surface.

The drug yields 6-15% oleo-resin consisting of mono-, bi-, tri-and tetracyclic phloroglucinol derivatives, which are formed by the condensation of a molecule of butyric acid with one (aspidinol), two (albaspidine) or three (filicic acid) molecules of phloroglucinol. The so-called 'filicin' is a mixture of ether soluble acidic compounds mostly the di-or tricylic phloroglucinols. Also present are aliphatic alcohols, triterpenoids and a flavone, dryopterin.

The drug is an effective taenicide against tapeworm. It is also used in cancerous tumours, as a contraceptive and a bactericide.

Semecarpus anacardium Linn. (Anacardiaceae)

Bhilawa, Marking Nut

This tree, a native of Asian and Australian tropics, possesses obovate leaves, small green flowers in terminal branched panicles and a sessile fruit with a black nut.

The nut contains a variety of phenols like bhilawanol, anacardic acid, cardol, catechol, anacardol, semecarpol and a fixed oil. Bhilawanol is a catechol derivative with a $C_{15}H_{37}$ side chain, and this chain, due to its lipid nature, helps the compounds to be absorbed through the skin. The resulting colour may be due to the oxidation of catechols to orthoquinones, which, in turn, get polymerized to coloured complexes.

The resinous juice extracted from the nut is used to remove rheumatic pains, aches, sprains and for any sort of veneral complaint or leprosy. In small doses, it is a stimulant and a narcotic.

Humulus lupulus Linn. (Cannabinaceae)

Hops

H. lupulus is a twining scabrid dioecious perennial herb with a stout branched rootstock, stem prickly with reversed bristles, large opposite cordate leaves (upper ovate, lower 3-5 lobed), male flowers in panicles, female flowers in pairs in the axils of broad bracts of a catkin-like ovoid spike and flattened achenes. Hops is native of N.America.

Bracts and flowers of female plants forming a leafy conical inflorescence (strobilus) collected before fully matured (contain glandular hairs from bracts) form the 'hops' of commerce.

Hops contain volatile oil and bitter phloroglucinol derivatives with isoprenoid side chains like lupulones (iso-α and β-acids, lupulin, principal component), humulinones (γ-acids) hulupones (Δ-acids), 4-deoxyhumulones, isohumulinones, lupoxes a & b, and lupdoxes A & B and polyphenols. Volatile oil contains thioester, terp-Me-sulphide and 2-Me-3-butene-2-ol.

Hops stimulates appetite, regulates menstrual period and is an antibiotic. It exhibits sedative and hypnotic effects and is a source of antioxidant phenols.

Capsicum annum Linn. (Solanaceae)

Syn. *C. frutescens* C.B.Clarke,
C. purpureum Roxb.
C. minimum Roxb.

Lal Mirch, Red Chilli

C. annum is a herbaceous perennial with oblong glabrous thin leaves, solitary white flowers from internodes, long green berries changing to orange to red containing pungent white/yellow circular seeds. A native of tropical America, this plant is now cultivated throughout the world. A number of varieties and cultivars are available.

Chillies are found to contain an oleoresin (16%), ascorbic acid, protein (16% dry wt.), fat (6%, mainly in seeds), carbohydrates (32%), carotenoids, a volatile oil and vitamin E. The oleoresin, yield varies from 15-18%; extracted by solvents, consists of capsainoids and some carotenoids. The most important component of capsainoids is a phenylamide capsaicin that is generally responsible for the pungency of chilly. Other capsainoids are dehydrocapsaicin, homocapsaicin II, bishomo-and trishomocapsaicin, homodihydrocapsaicin I and II, nordihydrocapsaicin, vanillyl amides of caprylic, nonylic and decylic acids and capsiamide. Capsaicin is maximum before the ripening stage of the fruit. Capsanthin and capsorubin, the carotenoids responsible for the red colour, are absent in green chillies, which contain lutein, neoxanthin, violaxanthin and β-carotene. Red pepper contains capsanthine-5, 6-epoxide, capsochrome and capsanthin 3,6-epoxide also along with β-carotene, violaxanthin, cryptoxanthin and cryptocapsin. (Lutein is absent in red peppers.) Vitamin C is more in large chillies. The volatile oil, which varies from 0.1 to 2.6%, consists of 2-methoxy-3-isobutylpyrazine, *trans*-β-ocimene, limonene, methyl salicylate, linalool, etc. Fruit pericarp yielded glycosides of luteolin, quercetin, and apigenin and p-coumaric and ferulic acids (Materska et al. 2003).

Seeds yield a fatty oil rich in linoleic acid and many steroids such as 4α-methyl-5α-cholest-8(14)-en-3β-ol, fucosterol, isofucosterol, citrostadienol and derivatives of lanosterol. The seeds also contain free sugars and capsicoside.

Chilli is a very popular spice and, in small doses, acts as a powerful stimulant and carminative. It is a well-known remedy in atonic and flatulent dyspepsia and dipsomania. A regular use is beneficial in varicose veins, anorexia and liver congestion.

Chilli is used as a prophylactic, irritant and rubifacient. It produces warmth, redness and burning sensation. The oleoresin is used (in the form of ointment of plasters) for the treatment of rheumatism, lumbago and neuralgia. Capsaicin is a potent stressor agent. Carotenoid extracts of this plant are found to be anti-inflammatory.

The green fruits contain many polyphenols (which are found to be absent in mature ones). They are petunidin, luteolin and quercetin glycosides, esters of *p*-coumaric and quinic acids, acyclic diterpene glycosides such as capsiansides I to V and A-E. All parts except the flowers and stem contain solanine and solanidine.

Curcuma longa Linn. (Zingiberaceae)

Haridra/Haldi, Turmeric

Turmeric is a native of China and E. Indies. This is cultivated everywhere for its rhizome, which is used as a spice. This plant is a perennial with a short thick rhizome (dark yellowish-brown inside), simple long-petioled ovate-lanceolate leaves and pale yellow flowers in dense bracteate strobiliform spikes.

The rhizome contains curcuminoids, volatile oil, sterols, sugars, starch and other polysaccharides. Curcuminoids are the principal colouring materials (upto 6%) of which curcumin amounts to 60%, with desmethoxy curcumin, *bis*-desmethoxy curcumin and dihydrocurcumin forming the rest. Volatile oil contains high amounts of bisabolene derivatives alongwith borneol, camphene, linalool and α-phellandrene. The polysaccharides reported are ukonans A-D. (Moon et al. 1977; Tomoda et al. 1990; Oshiro et al. 1990).

Curcuma shows excellent anti-inflammatory properties due to curcumin and its derivatives. It is also antihepatotoxic, antiulcer, anti cancerous, as also an antioxidant.

Piper methysticum Forst. (Piperaceae)

Kavakava

Kava consists of the dried rhizome and roots of a bushy dioecious shrub. This plant reaches a height of 2-3 m and is with thick knotty greyish green roots, cordate orbicular leaves, and monosporangiate flowers in the axils of peltate rounded floral bracts borne on pendulous spikes. Kava is indigenous to Fiji and nearby pacific islands.

The rhizome and roots, after removal of the bark, are cut into small pieces, chewed to make them fine and fibrous, and allowed to ferment in water to form the beverage. The drug yields 5-10% of a resin consisting of a number of styryl pyrones, such as dihydromethysticin, methysticin, yangonin, etc., and a considerable amount of starch. Pyridine alkaloids like pipermethystine, 3α-4α-epoxy-5β-pipermethystine and awaine are isolated from the aerial parts (Dragull et al. 2003).

The beverage possesses sedative and hypnotic properties, and is widely consumed by the natives. The pyrones are potent skeletal muscle relaxants and exhibit antipyretic and local anaesthetic properties.

A number of controversies exist on Kavakava. Since the active sedative components are kava lactones, standardized extracts of kavakava were prepared in Europe by extracting the roots in acetone and sold over the counters. These extracts lead to many serious health problems and even death. A recent study by Whilton et al., (2003) revealed that glutathione, which is also extracted in traditional aqueous preparations, reduces the toxicity of kavalactones by the Michael reaction in which the lactone ring is opened, rendering the kava extracts non-toxic.

Styrax spp. (Styraceae)

Benzoin

Benzoin is the balsamic resin obtained from various species of *Styrax* such as *S. benzoin* Dryand, *S. paralleloneurus* Perkins (Sumatra benzoin) and *S. tonkinensis* Craib (Siam benzoin). All these plants are native of S. E. Asia and East Indies.

Benzoin is a pathological product induced by the continuous incisions and subsequent fungal attack.

Sumatra benzoin occurs in the form of yellow irregular pebbles with a milky centre embedded in a reddish brown translucent resinous matrix and possesses agreeable odour and an acrid taste. Benzoin contains upto 20% free balasmic acids in which cinnamic acid amounts to 10% and benzoic acid 6%. Also present in the resin are the esters of benzoic and cinnamic acids, triterpene acids like 19-hydroxy oleanolic acid (siaresinolic acid), 6-hydroxy oleanolic acid (sumaresinolic acid), traces of vanillin and phenyl ethylene.

Siam benzoin occurs as concavo-convex yellow brown tears with a milky white centre embedded in a glossy reddish brown matrix and possesses a vanilla-like odour and balsamic taste. The important constituents of this resin are coniferyl benzoate (70%), benzoic acid (10%), a triterpene siaresinol (6%) and vanillin (in traces).

Benzoin is a stimulant, expectorant and used in the preparation of perfumes, soaps, toothpowders and cosmetic lotions.

4.6 Xanthones

Xanthones resemble benzophenones in having a C_6-C_1-C_6 skeleton but differ in the presence of a heterocylic oxygen ring in between the two benzene rings. More than 80 compounds are known of which almost all are hydroxylated (Carpenter et al. 1969). These compounds have a very restricted distribution and are similar to flavonoids in chromatographic behaviour and colour reactions. The Gentianaceae are found to be rich in these compounds. Gentisin is a yellow pigment obtained from *Gentiana lutea* Linn. roots. A xanthone C-glycoside, mangiferin, is reported from *Mangifera, Hedysarum, Madhuca* and *Iris* (Harborne, 1964).

4.7 Coumarins

Coumarins are formed by the lactonization of *o*-hydroxy cinnamic acid. More than 1200 coumarins are known from higher plants, especially dicotyledons. Coumarin itself is present as an aromatic constituent in many plants and in glycosidic form in *Melilotus*. Almost all the naturally occurring compounds are O-substituted at C_7. 7-Hydroxy coumarin—umbelliferone—is widely distributed in the Apiaceae while daphnetin (7,8-dihydroxy coumarin) is common in the Thymeliaceae and Euphorbiaceae. Mammein, with isoprenoid side chains (from *Mammea americana* Linn.), is a known insecticide. Scopoletin is the most common coumarin in higher plants.

Furano-, and pyranocoumarins have a furan or pyran ring fused with the benzene ring of coumarins. 6,7-Furocoumarins (psoralens) are abundant in the Umbelliferae and Rutaceae. Some of the furocoumarins are employed as fish poison while some others are effectively used in medicine for their spasmolytic and vasodilating effects. Bergaptene is the bergamot camphor and imperatorin (marmelosin) is a principal constituent of *Aegle marmelos* (fruit). Psoralen derivatives are administered orally to promote the susceptibility of suntanning of the skin (Musajo and Rodghiero, 1972). Isopsoralens are 7, 8-furocoumarins. Pyranocoumarins are also designated as chromanocoumarins.

Isocoumarins are a small group of α-pyrones reported from the Saxifragaceae and related groups. Bergenin is the most widespread member in this group. Phyllodulcin, a dihydroisocoumarin, is the sweetening agent in *Hydrangea*.

Aegle marmalos Correa (Rutaceae)

Bel, Bangal Quince

A. marmalos is an aromatic, moderate-sized tree armed with 2 straight stipular spines at the base of trifoliolate leaves (leaflets ovate lanceolate, lateral ones sessile and terminal, long petioled), large greenish white sweet-scented flowers borne in short axillary panicles and globose grey fruits having a woody rind enclosing a number of oblong compressed seeds embedded in a thick orange-colored sweet pulp. Bel is common in deciduous forests and cultivated near Hindu temples.

The fruits contain a group of coumarins, volatile oil, gum and tannins. The principal coumarins are furanoderivatives marmelosin and allo-imperatorin. Marmalide, psoralen, scoporone, scopoletin, umbelliferone, marmasin and skimmine are the other coumarins present. Also present is β-sitosterol. Tannin is more concentrated in the rind (18-22%). The gum is an acidic polysaccharide with galactose (20%) arabinose (10%) and galacturonic acid (25%). Seeds yield a fixed oil (35%) with oleic (30%) and linoleic acids (36%) as the major fatty acids. The leaves yield alkaloids aegeline and aegelinine, lupeol, sitosterol, rutin, marmesinin and β-sitosterol. The bark contains glycosides of lignans (-)-lyonoresinol, (-)-4-epilyoniresinol, (+)-lyonoresinol and a quinoline alkaloid skimmianine, β-sitosterol, marmin, umbelliferone and lupeol. Roots yield xanthotoxin, 6,7-diOMe coumarin scopoletin, umbelliferone, skimmmianine, skimmin, decursinol and haplopine (Yadav et al. 1989).

The ripe fruit is eaten fresh and the pulp is used for making juices. The unripe fruit is an astringent, digestive and stomachic and is effective in chronic diarrhoea and dysentery. It is also considered a tonic to heart and brain as well as hypoglycemic and anthelmintic. Roots, a component of *dasamoola*, are hypoglycemic and useful in hypochondriasis, melancholia and palpitation of heart. Leaves are useful in asthma and against worms.

Feronia limonia Swingle (Rutaceae)

Syn. *F. elephantum* Corr.

Kota, Wood Apple

F. limonia is a moderate-sized armed tree with axillary thorns, pinnate leaves containing obovate leaflets, white polygamous flowers in lax panicles and a large globose berry having a woody pericarp, and numerous compressed seeds embedded in an aromatic sweet pulp. A native of Srilanka, Kota is found throughout India.

The edible pulp of fruit contains protein (7%), carbohydrates (15%) and minerals. The leaves yield an essential oil (0.73%), containing estragol (90%) as the main constituent, a flavonol 3'OMe-quercetin and phenolic acids like p-hydroxy benzoic, vanillic, syringic, p-coumaric and ferulic acids. A gum, naturally exuding from bark, consists of arabinose, xylose and α-galactose.

The fruit is a tonic and used as a substitute for *Aegle marmelos*, for diarhhoea and dysentery. The gum is used a substitute for gum acacia.

Psoralea corylifolia Linn. (Fabaceae)

Bavchi, Psoralea

Psoralea, a native of India, is an erect herb, with roundish ovate leaves and violet flowers borne on spike-like dense racemes. The fruit is a one-seeded indehiscent legume.

Seeds yield about 1% furanocoumarins, psoralen and isopsoralen. Roots are found to contain an essential oil, chromonochalcones and isoflavonoids.

Seeds are used as stomachic, deobstruent, in leprosy and leucoderma. (Both psoralen and isopsoralan possess the curative property of *Psoralea* in leucoderma). Psoralen derivatives are taken orally to promote suntanning of the skin and are found to be spasmolytic and vasodilatory.

Eclipta alba Hassk. (Asteraceae)

Syn. *E. erecta* Linn.
E. prostrata Linn.

Bhringaraja/Bhangra

E. alba is an erect/prostrate annual herb, often rooting at nodes, with stems and branches covered by white hairs. Leaves sessile, oblong-lanceolate with appressed hairs on both sides, heads are heterogamous, white and achenes cuneate and winged. This medicinal plant is a very common weed in cultivated fields.

The plant yields (upto 1.6%) coumestan derivatives like wedelolactone and demethylwedelolactone. Thiophenes present are ecliptal(a terthienyl aldehyde), 2-angeloyloxy methylene-5'-(but-3-en-1-ynyl) dithiophene and

5-isovaleryloxy methylene-2-(4-isovaleryloxy-but-3-ynyl)dithiophene. Saponins (eclalbosaponins 1-1V), triterpenoids, flavone (luteolin), alkaloid (ecliptine) and polypeptides are other constituents of the plant. Roots are rich in thiophene acetylenes such as 5'-senecioyl oxymethylene-2-(4-isovaleryloxy but-3-ynyl)-dithiophene, 5'-tigloyloxymethylene-2-(isovaleryloxy but-3-ynyl)-dithiophene and 2-(3-acetoxy-4-chloro-but-1-ynyl)-5-(pent-1,3-diynyl)-thiophene (Sikroia et al. 1982; Wagner et al. 1986; Das and Chakravarthy,1991; Singh and Bhargava, 1992).

The drug is hepatoprotective (antihepatotoxic) due to wedelolactone and demethylwedelolactone. It is also antimyotoxic and antihaemorrhagic in nature.

4.8 Chromones

Chromones resemble coumarins in the overall structure and properties but differ in the biosynthesis and position of the ketogroup. Chromones are formed by acetic acid pathway (by the condensation of five acetate molecules) while coumarins are derived from shikimic acid pathway. About a dozen simple chromones are known to occur naturally. These compounds, generally, have a methyl group at C_2 and are oxygenated at C_5 and C_7. An example of simple chromone is eugenin (in *Eugenia caryophyllata).* Khellin, visangin and khellol are some furochromones obtained from *Ammi visnaga* Lam. and *Eranthis hyemalis* Salisb, having certain marked pharmacological activity. Khellin is a powerful vasodilator and antispasmodic.

Flavones and isoflavones are 2-, and 3-phenyl chromones, respectively, whereas rotenone is a complex chromone derivative.

Bergenia ciliata Sternb. (Saxifragaceae)

Syn. *B. ligulata* Engl.

Pashanbheda, Megaseas

This is a perennial herb with a stout underground rhizome, extremely variable broadly obovate denticulate leaves, small white flowers in cymose panicles and spherical capsules containing many small arillate seeds. A native of Himalayas, *B. ciliata* is common in the hilly areas of north India.

The dried rhizome, which forms the drug, yields an isocoumarin deriv. bergenin (0.6%), gallic acid, flavonoids, tannins (afzelachin), glucose and β-sitosterol.

Pashanbheda is considered to be very effective in dissolving kidney stones. The drug is a tonic, antiscorbutic, and is given for pulmonary infections, ulcers, cough and fever. It is also found useful in vertigo and headache.

Ammi visnaga Lam. (Apiaceae)

Visnaga, Khella

A. visnaga is a tall robust herb with a thick stout stem, deltoid pinnately decompound leaves, white flowers in dense umbels and oblong ovoid fruits resembling caraway. Visnaga is a native of the Middle East.

The fruits yield 0.5–1.5% chromones consisting of khellin, visnagin, khellinin, visnaginol and visinagionol and a number of furanocoumarins such as bergaptene, marmelosin, marmesin and xanthotoxin. Protein (12%), fat (18%) and a volatile oil are the other consitituments of fruits. The leaves and flowers contain flavonols (myricetin, quercetin and kaempferol alongwith their glycosides), whereas all parts of the plant are found to contain khellin, visnagin and khellinin.

Fruits are used for expelling kidney stones and for renal colic. Khellin is antispasmodic and a vasoditator. It is also used in the treatment of angina pectoris and asthma.

4.9 Stilbenes

Stilbenes are C_6-C_2-C_6 compounds, about whom nine are known, occurring mainly as heartwood constituents. Most of these compounds are toxic to fungi, fish and insects. Some of the stilbene derivatives are the condensed tannins. Rhapontigenin from rhubarb, pterostilbene from *Santalum* and pinosylvin from *Pinus* are some of the interesting members of this group.

Lunularic acid is a dihydrostilbene found in lower plants, functioning as a growth inhibitor analogous to abscisic acid.

4.10 Lignans

Lignans are dimers formed by the condensation of two cinnamic acids/ cinnamyl alcohols, through the β-carbons of their aliphatic side chains. The aromatic rings are often oxygenated and additional rings are also formed in the molecule. Optical activity is exhibited by many lignans. Once considered as typical heartwood constituents associated with lignin (so the name lignans), these compounds are located in leaves, fruits and in exudates of many plants. Very few glycosides are reported from this group of which about 85 members are known to exist naturally in higher plants. Lignans are extensively used as antioxidants in food. The stability of sesame oil is attributed to the lignan, sesamin, present in it. The same lignan forms the synergistic ingredient in pyrethrum insecticides. Podophyllotoxin, the principal component of *Podophyllum* resin along with α-and β-peltatins, exhibits powerful cathartic action. Podophyllotoxin, being cytotoxic, is found useful in treating human malignancies (neoplasms). A number of lignans are effective antimicrobial agents.

Lyonia–xyloside occuring in the sapwood of *Lyonia* (Ericaceae) and arctiin (arctigenin glucoside) in the fruit of *Arctium* (Asteraceae) are two examples of lignan glycosides. Though lignans are considered intermediates in the formation of lignin, they are also produced as a result of fungal attack and thus serve as phytoalexins.

The lignans derived from propenyl phenols or allyl phenols are designated as neolignans, of which about 130 members are known. Otobain, present in the otoba butter *Myristica otoba* H&B. fruit, is used as an antifungal factor. Schizandrin is one of the active principles of *Schizandra chinensis* C. Kech. used in the Orient as a stimulant, while nor-dihydroguaiaretic acid and *l*-guaiaretic acid from *Guaiacum* Linn. is extensively used as an antioxidant in lipids. Nor-dihydroguaiaretic acid is described as the most potent anticancer metabolite in vitro (Burks and Woods, 1963).

Trimeric and tetrameric lignans have also been reported recently in nature (Wagner et al. 1975). These compounds, on detailed analysis, would help immensely in our understanding of the complexity of lignin and related polyphenols.

Podophyllum peltatum Linn. (Berberidaceae)

Mandrake/May-apple

Mandrake is the dried rhizome and roots of *P. peltatum.*, a native of Canada and the US. It is an erect herbaceous perennial with horizontal rootstock and large peltate leaves. The leaves are palmately 5-9 lobed, lobes usually 2-cleft and dentate at the apex. The nodding, white, solitary large flowers are usually borne on the fork between two leaves. The fruit is a fleshy yellow berry.

The rhizome and roots yield 2-8% of a resin consisting of a group of lignans. The three principal lignans podophyllotoxin, α-, and β-peltatins amount to 20%, 10% and 5%, respectively. These compounds occur as glycosides or as aglycones. Also present in the resin are demethylpodophyllotoxin, dehydropodophyllotoxin, quercetin and starch.

Mandrake is a drastic purgative. Podophyllotoxin shows excellent antimitotic action.

Podophyllum emodi Wall.

Bakrachimyaka, Indian podophyllum

This perennial herb of the Himalayas yields 6-12% resin with a higher percentage of podophyllotoxin (40%). Excepting the peltatins, the other ingredients are the same as in *P. peltatum*.

Podophyllotoxin—along with the other lignans, α-, & β-peltatins and 3'-demethyl podophyllotoxin—is responsible for the anti-tumour activity exhibited by *Linum album* Kotschy (Weiss et al. 1975).

Phyllanthus amarus Schum. & Thonn. (Euphorbiaceae)

Syn. *Phyllanthus niruri* Linn.

Bhoomyamalki/Bhuiavla, Phyllanthus

This is a small annual herb with small ovate/elliptic leaves distichously arranged on branches (resembling a pinnate leaf, similar to emblica), small unisexual flowers (female flowers towards the base of branches and male flowers towards the tip) and spherical 6-lobed capsules containing ribbed seeds. *P. amarus* occurs as a weed throughout.

The entire plant, which forms the drug, contains lignans phyllanthin (a diaryl butane, major constituent, 0.5%) and, hypophyllanthin (aryltetrahydronaphthalene, upto 0.2%) as major constituents. Nyrphylline (lignan), phyllnirubin (a neolignan), hydrolyzable tannins such as phyllanthusin D, amariin, amarulone and amarinic acid and alkaloids like *ent*-norsecurinine, sobubbialine and epibubbialine are the other constituents of the drug.

P. amarus is known for the hepatoprotective action it exerts, an action due to the lignans. The drug exhibits antiviral actions on hepatitis B. It is also used as hypoglycemic, diuretic and hypotensive.

Gmelina arborea Roxb. (Verbenaceae)

Kasmari, Gmelina

This is a moderate-sized deciduous tree with a greyish-yellow corky bark, pubescent large broadly ovate long-petioled leaves, brownish yellow flowers in groups of 3 on a hairy panicle and an ovoid drupe containing 1-4 seeds. Kasmari is common in India, Malaya etc.

The roots of this tree contain gmelofuran (a cadinane type furanosesquiterpene), β-sitosterol, gmelinol, hentriacontanol, ceryl alcohol and n-octacosanol. Heartwood yields ceryl alcohol ,β-sitosterol, gmelinol, cluytyl ferulate and lignans such as arborone, 7-oxodihydrogmelinols, arboreal, gmelanone, 6"-bromoisoarboreol, epieudesmin and gummidiol. Leaves contain alkaloids, flavonoids like apigenin, luteolin, quercetin and quercetogenin besides β-sitosterol and hentriacontanol.

The wood is hypoglycaemic and also used in skeletal fractures. The various parts are used in fever, dropsy, anasarca, spleen troubles, rheumatism, delirium, smallpox, syphilis, asthma and diarrhoea.

Piper cubeba Linn. (Piperaceae)

Kabab Chini, Cubeb

Cubeb is a native of E. India and Malaya. Its dried unripe fruits are used as a spice. These woody climbers are cultivated in Java, Thailand, Sri Lanka and W. Indies. The spikes bear more fruits than those of pepper, and the berries have 'false' stalks due to the abnormal development of the pericarp.

Cubeb contains lignans, 7.5% resins, 8% gum, 1% fixed oil and 18% volatile oil. The lignans present are cubebin (0.8%), hinokinin, clusin, dihydrocubebin, cubebinolide, cubebinone and 5" methoxyhinokinin. Volatile oil comprises of sabinene, Δ^4 carene, 1-4 cineole, sesquiterpenes like *l*-cadinene and alcohols. Also present are oxygenated cyclohexanes such as piperonal A & B, crotepoxide and zeylanol and sesquiterpenes bicyclosesquiphellandrene and *l*-epibicyclosesquiphellandrene.

Cubeb is a carminative, antiseptic and is used medicinally for urinogenital diseases.

Sesamum indicum Linn. (Pedaliaceae)

Til, Sesame

Sesame is a native of India, cultivated extensively in almost all tropical countries like China, India, Africa and Latin America. The plant reaches a height of 1m, with ovate-lanceolate leaves, pale rose or white flowers singly in the axils of leaves and a grooved capsule containing numerous small, flattened, reddish-brown seeds. The oil is obtained by cold pressing these seeds.

The seeds yield about 50% fixed oil, 22-25% proteins and upto 8% mucilage. The oil, a pale yellow odourless and tasteless liquid, consists of 45% oleic, 40% linoleic, 10% palmitic, 4% stearic and traces of arachidic and myristic acid glycerides. Also present in the oil is a phenol, sesamol, produced by the hydrolysis of a lignan, sesamolin.

Sesame oil is found be one of the stablest oils; the stability attributed to the antioxidant properties of sesamol. It is used as edible oil, in the preparation of margarines and possesses laxative and demulcent properties. The poorer grades of oil are used for making soaps, lubricants and in perfumery. Sesame seeds are used as food, oil cake as cattle feed and sesamolin as an insecticide.

Linum usitatissimum Linn. (Linaceae)

Flax, Linseed

Linseed is the dry ripe seed of flax, a fibre plant. This plant, a native of S.E. Asia, is a slender branching annual with small linear-lanceolate leaves, small blue/white flowers borne on terminal leafy panicles and a globular capsule containing small ovate flat seeds.

The seed of flax contains fixed oil (37%), protein (20%), carbohydrates (29%) and phosphatides (lecithin and cephaelin, 0.25%). Mucilage (amounting to 2-7%, concentrated in the hulls) and a cyanogenetic glycoside linamarin are the other constituents. The hull enclosing the seed is rich in lignans such as secoisolariciresinol diglycoside (SDG).

Linseed is a demulcent, emollient, expectorant, and diuretic. The entire seed is prescribed as a laxative, like isabgul. As an infusion in water, linseed tea is used in fever, cold, urinary infections, gonorrhoea and diarrhoea. Mucilage is used for conjunctivitis as an external application. The lignans present are useful in prostrate and skin cancer, melanoma and diabetes. They also act as phytoestrogens useful in treating breast tenderness, sweating, vaginal dryness, hot flushes, uneven menstrual cycle and PMS symptoms. Lignans are produced by intestinal bacteria also from precursors present in linseed meal.

Schisandra chinensis Linn. (Schisandraceae)

Magnolia Wine, Wu-wei-zi

Schisandra is an aromatic woody wine with simple toothed leaves (often pellucid dotted in spiral arrangement, pink unisexual flowers (flower with many stamens and carpels spirally arranged on an elongated receptacle) in a few flowered inflorescences in axils of leaves and an etaerio of red berries each containing two seeds. The fruits possess five basic flavours—salty, sweet, sour, pungent and bitter. Magnolia wine is a native of China.

The fruits contain lignans such as schisandrin, deoxyschisandrin, gomisin and pregomisin, essential oil and numerous acids.

In Chinese medicine, the berries are used as a tonic, astringent and expectorant and for hepatitis and diarrhoea. The other uses include as a kidney tonic, for physical exhaustion and insomnia and as a chemotherapy support. The lignans in *Schisandra* are found to regenerate liver tissue damaged by hepatitis, alcohol and other disorders.

Aleurites moluccana Willd (Euphorbiaceae)

Syn. *A. triloba* J. R. & G. Forst.

Candle Nut Tee

A.moluccana is a handsome evergreen tree of Indo-Malaysian origin, having long pendulous branches bearing stellate-pubescent 3-5 lobed large ovate leaves, small white flowers in tomentose cymes, subglobose fleshy 4-angled fruits containing one or two large hard seeds. This has been naturalized in India.

Stem chips are found to contain a coumarinolignoid, moluccanin. The hard shell constitutes 65-70% of the seed and kernels, 30-35%. Kernels yield 60-70% oil containing linoleic (49%), linolenic (30%) and oleic acids (10%) (Shamsuddin et al. 1988).

The oil is a drying oil used in paints and varnishes. Medicinally, it is used as a purgative and externally in rheumatism and ulcers. The leaves are applied in case of acute rheumatism. The fruit is considered as a tonic, used in diseases of heart and blood. It shows anthelmintic, carminative and expectorant properties and is used against piles and hydrophobia. The flowers as well as the bark are used against asthma, ulcers, and tumours and as a laxative.

4.11 Flavonoids

The term flavonoid, in the wider sense as it is used here, includes not only the compounds based on flavone (2-phenyl chromone) skeleton, but also a number of related or derived compounds like anthocyanins, isoflavones, neoflavones, etc. The different classes of these C_6-C_3-C_6 compounds are recognized chiefly by the oxidation pattern of the C_3 fragment, the additional oxygen heterocylic rings and the glycosylation. The spectra of properties exhibited by these compounds include colouring of flowers and fruits (anthocyanins, flavone/ols), as accessory pigments which absorb and pass on the solar energy from lower wavelengths to chlorophyll in leaves (flavones), astringency (flavonones) estrogenic (isoflavones) and so on.

Flavonoids often occur as glycosides in plants, but a number of them are present in the free state too. The sugars involved in glycosidic linkages are rarely unusual ones. C-glycosides also are not uncommon in this group. The various classes of flavonoids discussed here are:

1. Anthocyanins
2. Aurones
3. Biflavonyls
4. Chalcones
5. Flavan-3-ols, 3, 4-diols and proanthocyanidins
6. Flavones and flavonols
7. Flavonones, and flavononols
8. Isoflavones, isoflavonones and homoisoflavones
9. Neoflavonoids.

4.11a Anthocyanins

Anthocyanins are glycosides of anthocyanidins-derivatives of 2-phenyl benzopyrilium. These compounds are responsible for the various shades of red, purple, pink and blue colours of flowers and fruits. In general, they are red in acid solutions but change into blue in basic solutions. They occur as cations in the vaculor sap and are thought to be oxonium compounds with the positive charge residing on the heterocyclic oxygen. The quinonoid form, resulting by the action of alkali, gets rapidly oxidized in air, resulting in the formation of a colourless derivative. This explains the instability of anthocyanins in alkaline solutions.

About 16 anthocyanidins are known to occur in nature. Most of them occur as glycosides while a few are products of hydrolysis of proanthocyanidins (known earlier as 'leucoanthocyanins'). It is found that the increase in the number of methyl groups in a compound is the cause of redness and increase in the –OH groups or 5-glycosylation causes blueness. The association of metals and/or pH markedly alters the colour of an anthocyanin. The 3-position is frequently oxygenated. The preferred position of glycosylation is C_3 for monosides and 3 & 5 for diglucosides (Among flavonoids only anthocyanins are glycosylated simultaneously in two positions.) In acylated anthocyanins, cinnamic acids like *p*-coumaric, ferulic and caffeic acids are seen linked to the sugar component, which in turn, are linked at C_3 position of the anthocyanidins.

The common anthocyanidins are cyanidin, delphinidin and pelargonidin. These compounds differ in the hydroxylation pattern of the B ring. Petunidin, malvidin and penoidin are the mono/dimethoxy derivatives of these compounds. The two 3-deoxy anthocyanidins apigenidin and lutelinidin are characteristically present in ferns and mosses (Crowden and Jarman, 1974). The anthocyanins of fruits like blueberry are found to delay cataract and related problems of the eye and increase night vision. The red anthocyanins of wine are considered a reason for the low incidence of cardiac problems in French people who are known to consume large quantities of this alcoholic drink.

4.11b Aurones

Aurones are benzal coumaranones. Alongwith related chalcones, these compounds are known as 'anthochlor pigments' due to their bright yellow/orange colour. Aurones possess a furan ring instead of the pyran ring of other flavonoids. They are reported from flowers, bark, wood and leaves. The known aglycones, seven in number, are mostly confined to the advanced families like the Cyperaceae and Asteraceae. The most widespread compound is aureusidin.

4.11c Biflavonyls

Biflavonyls are dimeric compounds in which two flavones are linked through a C-C linkage (amentoflavone) or an oxygen bridge (hinokiflavone). Biflavones are widely distributed in the gymnosperms and are rare in Angiosperms. Almost all the known compounds, about sixty, are dimers of apigenin. Morelloflavone contains a flavone and flavonone units.

4.11d Catechins, Flavan-3, 4-diols and Proanthocyanidins

All these compounds are devoid of the keto group at C_4 and possess a saturated C_3 fragment, making them 'flavans' instead of flavones. They are optically active (the activity due to the asymmetric nature of C_2 and C_3), colourless compounds omnipresent in higher plants and concentrated especially in woody plants. Catechins are flavan-3-ols occurring free in certain gums or bound with flavan –3, 4-diols in proanthocyanidins. Depending upon the relative positions of the two H-atoms at C_2 and C_3, the isomer of catechin, epicatechin is recognized. Both (+) and (-) catechins contain 2 and 3 hydrogens *trans,* whereas in epicatechins they are in *cis* form.

Flavan-3, 4-diols are components of proanthocyanidins. A proanthocyanidin molecule may be a dimer of flavan 3, 4-diol or contain a catechin and a flavan-3, 4-diol. These compounds, as the name signifies, yield anthocyanidins (25%) on acidic or alkaline hydroloysis along with complex polymeric structures known as 'phlobaphenes' or 'tannin reds'. The polymeric phlobaphenes and the proanthocyanidins form the condensed tannins. All these compounds are termed flavolans. Proanthocyanidins seldom occur as glycosides. The linkage between the two monomers, most often, is a C-C linkage. The beneficial role of proanthocyanins in preventing heart-related problems is well known.

4.11e Chalcones

Chalcones are yellow pigments possessing an open chain system for the middle C_3 fragment. The numbering of carbon atoms also is different from other flavonoids. About two dozen chalcones are known to occur in plants. Reported even from primitive vascular plants like ferns, they are considered very early in evolution. The labelling experiments prove them to be the first C_{15} precursors formed in flavonoid synthesis. In acid solution, chalcones are converted to flavonones while the reverse reaction occurs in alkaline medium. Thus, the butein (chalcone)-butin (flavonone) system is inter-convertible.

A hydroxyl group is always present in position-2, corresponding to the hetero oxygen atom of flavones. Phlorizin, (the 6-glucoside of phloretin-2, 4, 6, 4'-tetrahydroxy dihydrochalcone) occurring in *Malus*, causes glycosurea in animals. Carthamin and isoliquiritigenin glycosides are the colouring principles of safflower and liquorice, respectively. Angolensin, obtained from the heartwood of *Pterocarpus angolensis*, is the only example of an iso-chalcone.

4.11f Flavones and Flavonols

These compounds, together known as anthoxanthins (yellow flower pigments), are by far, the most abundant group of flavonoids. Most of them are yellow in colour and are responsible for the colour of many flowers ranging from yellow to white. When they co-occur with anthocyanins, they exert a blueing effect on the flower colour. Flavone itself occurs in primrose flowers and its various hydroxylated and methoxylated derivatives contribute to the light yellow or white colour of many flowers. Apigenin, luteolin and their methoxy derivatives are widespread in plant kingdom. The glycosylation normally occurs at C_7, and the sugar component may be a single sugar like glucose (most frequent), galactose or rhamnose or a disaccharide (contrary to the anthocyanin diglucosides; two sugars are never attached to different hydroxyl group in a flavone). The glycosides are present dissolved in cell sap while the free aglycones occur in the secretion products.

Flavonols are 3-hydroxy flavones. They are darker in colour and contribute to the deep yellow colour of many flowers. Excepting that the glycosylation is common in C_3, the properties of these compounds are very similar to flavones. The three common members of this group are kaempferol, quercetin and myricetin. Rutin (vitamin. P) is the 3-O-rutinoside of quercetin. Quercetin is considered an anticancerous drug which exhibits antihistamine and antiinflammatory properties also.

4.11g Flavonones and Flavononols (Dihydroflavones and Dihydro Flavonols)

These compounds, though colourless, occur in appreciable quantities in several economically important plant products. Similar to flavans, they contain a saturated C_3 fragment, which renders them colourless. About 80 flavonones and 40 flavononols are known today.

These compounds are optically active and a majority of them are *laevo* rotatory. Flavonone with only one asymmetric carbon (C_2) can exist in two forms while flavononols (3-hydroxy flavonones) with two asymmetric carbons (C_2 and C_3) are capable of existing in four forms.

Flavonones are stable in acidic medium but are converted to chalcones in warm alkali solutions. The flavonones corresponding to apigenin and luteolin, naringenin and eriodictyol are widely distributed among higher plants, but their low concentration is the reason for their not being detected in a routine survey. Flavononols are a much neglected group, of which very little is known. Taxifolin (dihydroquercetin) is the most common compound in this group.

Hesperidin and naringin, two glycosides of the flavonones hesperetin and naringenin, are responsible for the bitter taste of the juice and peel of *Citrus* fruits. Both flavonones and flavononols are fungitoxic compounds useful in wood preservation. Both these groups of compounds are considered obligate intermediates in flavonoid biosynthesis. Despite the similarity with flavans, no condensed tannins are known from this group.

Silymarin from *Silybium* is a group of flavonolignans (in which a flavone skeleton is linked with a phenylpropane group) having marked antihepatotoxic activity.

4.11h Isoflavones, Isoflavonones and Homoisoflavones

Isoflavones are 3-phenyl chromones. They are isomeric with flavones and differ in the placement of B ring to the C_3 position. The hydroxylation pattern also is similar to flavones. Genistein and orobol correspond to apigenin and luteolin. The total number of isoflavones isolated exceeds 50.

Isoflavones are known to be the heartwood constituents of Fabaceae plants where they are particularly abundant in the subfamily Lotoideae. These compounds are of great interest, eliciting diverse physiological responses. 5,

7-Dihydroxy isoflavone is a strong inhibitor of oxidative phosphorylation. Pisatin and trifolirhizin are produced in response to injury or fungal attack and are protective in nature (Phytoalexins). Genistein and pterocarpans are estrogenic and fungicidal. Genistein and diadzin are two isoflavonoids widely used in treating and preventing breast cancer. Rotenone is a known insecticide.

The different structural variations found in isoflavones include: (1) *isoflavonones* (e.g.: Padmakastein from bark of *Prunus puddum* Roxb.); (2) *rotenones*, with a 4-ring chromanochromanone system as in *Derris;* (3) pterocarpans having a coumaranochroman ring; (4) *reduced isoflavans* e.g. duratin from *Dalbergia;* (5) *coumestans* with highly oxidized coumaranocoumarin structure, e.g. coumestrol; and (6) 3-aryl, 4-hydroxy coumarins, e.g. lonchocarpic acid from *Derris.*

Homoisoflavones are higher homologues (C_{16}) of isoflavones. Only two compounds are known, both from *Eucomis bicolor* Baker (Liliaceae).

4.11i Neoflavanoids

This is a newly recognized group, characterized by a 4-phenyl chroman skeleton. The structural variations in this group include: (1) *4-phenyl coumarins* like calophyllolide obtained from Guttiferae members like *Calophyllum, Mesua* and *Mammea;* (2) *dalbergiones*, a group of benzoquinones isolated from *Dalbergia, Machaerium* (Fabaceae) and *Exostemma* (Rubiaceae); and (3) *aryl chromans* like brazilin and haematoxylin. Calophyllolide and related compounds are piscicidal while brazilin and haematoxylin are dyes.

Vaccinium myrtillus Linn. (Vacciniaceae/Ericaceae)

Huckleberry, Bilberry

Bilberry is a small branched shrub with wiry angular branches, leaves first rosy, turning yellow and finally red and leathery, globular wax-like flowers and black globular (having a flat top) berries containing many small red seeds of a sharp sweet taste. Huckleberry is a native of northern Europe and north America.

Bilberry contains at least 15 anthocyanins, flavonoids such as quercitrin, isoquercitrin, hyperoside and astragalin, catechol tannins, glycoquinic acid, ursolic acid, phenolic acids (caffeic and chlorogenic acids) and pectins. The anthocyanins are based on malvidin, cyanidin and delphinidin. The anthocyanin content increases as the fruit ripens mainly due to the conversion of proanthocyanidins and catechol tannins to anthocyanins.

Bilberry anthocyanins are used for improving eyesight as well as increased night vision, delayed cataract and other eye disorders. This is especially useful for age-related macular degradation (AMD), which is the cause of blindness in people over the age 55 and regenerate rhodopsin.

Crataegus laevigata DC & *C. monogyna* Linn. (Rosaceae)

Syn. *C. oxycantha* L. (for *C. laevigata*)

Hawthorn

Hawthorn is a moderate-sized deciduous ornamental tree with stout spines, 3-5 apically lobed leaves (having serrate margin), small white flowers in clusters (anthers many, red) and a scarlet subglobose fruit containing 2 nutlets which are furrowed on the inner side. A native of Europe, hawthorn is widely cultivated for the flowers.

Berries contain anthocyanins, flavonoids such as rutin, vitexin and orientin, proanthocyanins, cyanogenetic glycosides phenylethylamine, trimethylamine, tyramine, histamine, choline, acetylcholine, chlorogenic acid and fruit acids (citric, tartaric and crataegus acids). Leaves yield vitexin, hyperoside, proanthocyanins, chlorogenic acid, and oxalic acid and triterpene acids like crataegolic, acantholic and neotaegolic acids. Flowers contain a volatile oil rich in 3-pyridine carboxy aldehyde, flavonols and anthocyanins.

Hawthorn (berries, leaves and flowers) is taken orally to relieve chronic heart conditions, primarily congestive heart failure, angina and arrhythmia. It is also used for treating anxiety and insomnia.

Sambucus nigra Linn. (Caprifoliaceae)

Elderberry

A native of Europe, elderberry is a tall tree with oval leaves, cream-coloured flowers in terminal panicles and blue-black berries.

Flowers are found to contain flavonoids (rutin, isoquercitrin and kaempferol), triterpenes, steroids, volatile oil and tannins. Leaves yield cyanogenic glycosides and fruits contain anthocyanins and flavonoids besides vitamins A and C.

Berries are diaphoretic, diuretic and laxative. Leaves and flowers also possess diaphoretic property. All these parts, especially the berry are recommended for colds, fevers, chronic nasal catarrah, sinusitis and rheumatism. Elderberry is especially recommended for flu viruses.

Hibiscus rosa-sinensis Linn. (Malvaceae)

Japa, Shoe flower

Japa is an arborescent mucilaginous shrub with long-petioled, ovate-lanceolate leaves, serrated towards the tip, basally three-veined, large axillary solitary red flowers (with epicalyx, staminal tube and reniform anthers.). A native of China, this plant is now cultivated everywhere.

Flowers contain mucilage, anthocyanins based on malvidin and cyanidin and flavonoids such as quercetin and gossypetin. Leaves yield taraxeryl acetate, β-sitosterol and flavonoids such as acacetin, 4'-OMe vitexin, 7-OMe vitexin and phenolic acids vanillic, syringic, *p*-coumaric and ferulic acids (Sheeja,1991).

Leaves, stem and roots are used in treating gonorrhoea and stomach troubles. Flowers are emmenagogue and useful in treating vaginal and uterine discharges. They are effective in treatment of arterial hypertension and have significant antifertility effect. Flower extract is both cytostatic and cytotoxic. Flowers are also considered hypoglycemic, aphrodisiac and for piles. Leaves and flowers are good for healing ulcers and for promoting the growth and colour of hair.

Ixora coccinea Linn.(Rubiaceae)

Paranti, Ixora

This is a glabrous shrub with obovate, opposite leaves having a cordate base, interpetiolar cuspidate stipules, bright scarlet tetramerous flowers in terminal corymbose cymes and globose red berries containing 2 ventrally concave seeds. A native of India, Ixora is widely cultivated in gardens.

The leaves yield flavonols such as kaempferol and quercetin, proanthocyanidins, and phenolic acids like gentisic, protocatechuic, vanillic, syringic, melilotic, *p*-coumaric and ferulic acids. The flowers contain cyanidin and flavonoids (Thomas, 1989).

Ixora is a good blood purifier and used for skin ailments, especially for children. Leaves and stem are astringent, antiseptic and sedative. Roots are useful in diarrhoea, dysentery, gonorrhoea and fever. Flowers form a component of hair oil.

Fagopyrum esculentum Moench. (Polygonaceae)

Kotu, Buckwheat

Buckwheat is an annual of 1-2 m in height with alternate hastate leaves, small pinkish-white fragrant flowers in axillary or terminal cymes and a brown-grey 3-cornered achene having keeled edges. It is a native of Central Asia, grown or found in the wild.

The grain, considered as a pseudocereal, contains carbohydratres (65%) protein (10%), fat (2.5%), minerals of vitamin B. The leaves and blossoms yield rutin (3-5%) and fat, rich in lecithin.

Rutin is used in the treatment of increased capillary fragility with associated hypertension sometimes resulting in the bursting of blood vessels in the brain, retina, etc. It also prevents weakening of capillaries caused by the administration of drugs such as salicylate, arsenical, sulphadiazine, etc.

Camellia sinensis Kuntze (Theaceae)

Chai, Tea

C. sinensis is a hardy, much-branched tree (shrub is cultivation) with stalked elliptic serrulate leaves, light yellow flowers singly or in pairs, and tri or bicoccate fruits containing a suborbicular/hemispherical large seed in each coccus. A native of China, tea now cultivated in many tropical countries.

The young leaves, which form the popular beverage tea after processing, contain purines, polyphenols, saponins, aminoacids, coumarins and a volatile oil. Caffeine, the most important constituent of the tea, is a purine often accompanied by theophylline, theobromine, xanthine and adenine. Caffeine amounts to 4% in fresh leaves and buds. The polyphenols amount to 30%, of which about three quarters are flavanols generally referred to as catechin. The members of this group are (+)-and (-)-catechin, (+)-and (-)-gallocatechin, (-)-gallocatechin gallate, (-)-epiafzelchin, (-) – epigallocatechin and its gallate and epicatechin and its gallate. Of these, epigallocatechin and its gallate are the prinicipal components which, during fermentation, produce the pigments characteristic to black tea. Dimeric gallates of procyanidin and prodelphinidin, dimeric flavan–3-ol gallates (theasinensins A and B) and acylated flavan–3-ols, [(-) epigallocatechin 3-0-*p*-coumaroate, (-)-epigallocatechin 3,3'-di-O-gallate and (-) epigallocatechin 3,4'-di-O-gallate] and tropolones are the other members of catechins. Flavonols such as kaempferol, quercetin, and myricetin in glycosidic forms, saponaretin, vitexin, isovitexin, theiferins A & B and several other flavone glycosides are the other polyphenolic constituents of tea. Other phenolics of tea are chlorogenic, *p*-coumaryl quinic, gallic and ellagic acids and coumarins such as umbelliferone, skimmin and scopoletin. The saponins reported are theafolisaponin (based on barringtenol), α-spinasterol gentiobioside and the steroids present are typhasterol, teasterone, spinasterol and stigamastenol. 2-Amino-5-(N-ethyl carboxymido)-pentanoic acid, putresine, theanine, S-methyl metheonine are the amino compounds present in tea. In addition, tea contains organic acids (malic and oxalic acids), sugars (arabinose, galactose and xylose). A volatile oil consisting of α-murolene, δ-cadinene, linalool, geraniol, etc., is responsible for the aroma of green tea. The flavour of black tea is due to a mixture of dimethyl suplhide, acetaldehyde, propionic aldehyde, isobutyraldehyde, acrolein etc

Black tea, due to the fermentation during processing, contains polymeric flavonoid complexes theaflavins and thearubigins. Theaflavins are yellow compounds. The theaflavins are astringent compounds and polymerize further to tannin-like red thearubigins. Thearubigins are found to contain more than 5 flavan-3-ols/flavon-3-ol gallates with interflavanoid links as well as benzotropolone units.

Tea seeds yield fats (23%), protein (8.5%), starch (33%), and saponins (9%). The oil is rich is oleic (58%), linoleic (18%) and palmitic (18%) acids. The saponins present are based on sapogenins, theasapogenols A-E and α-spinasterol. Two phytohemagglutinins also have been isolated from the seed.

Tea is a stimulating beverage relieving muscular and mental fatigue due to caffeine. Caffeine is diuretic and smooth muscle relaxant. Tea polyphenols inhibit platelet aggregation and possess anticoagulant activity. They strengthen the walls of capillaries and are antiinflammatory. Green tea is found to inhibit the growth of cancer tumours in alimentary canal. Quercetin is a known antitumour agent. The fluoride present in tea prevents tooth decay. Catechin is useful in treating nephritis and chronic hepatitis. S-methyl methionine is Vit. U, the anti-ulcer factor.

Oroxylum indicum Vent. (Bignoniaceae)

Syn. *Bignonia indica* Linn.

Shyonaka, Oroxylum

O. indicum is a small tree, with a soft bark having green juice and corky lenticels. Leaves are very large (1-2m long) opposite, 2-3 pinnate; leaflets are large, ovate-elliptic and unequal; flowers are purple, large, zygomorphic with 5 stamens, 4 in didynamous condition; and the fruit is a long flat woody capsule containing many flat seeds, winged all around. Oroxylum is a common tree in India.

The stem and root bark contain three flavones, oroxylin-A (5,7-diOH,6-OMe flavone), baicalein (5,6,7-triOH flavone) and chrysin (5,7-diOH flavone). Also present are sitosterol and tannins. Seeds yield a fatty oil (of oleic 80% and rest saturated acids), baicalein and its glucoside tetuin.

One of the *Dasamula* (formulation of 10 roots, used as a tonic), the root bark is useful in diarrhoea and dysentery. Bark is used in acute rheumatism and is diaphoretic. It is also useful in skin diseases, oedema, urinogenital diseases, piles and cough. Leaves are useful in rheumatism and as a poultice for enlarged spleen, headache and ulcers.

Biophytum sensitivum DC. (Oxalidaceae)

Syn. *Oxalis sensitiva* **Linn.**

Lajjalu, Biophytum

B. sensitivum is a small herb with paripinnate leaves (leaflets 3-12 pairs), normally forming a rosette at the base, yellow flowers in terminal racemes and elliptic shining capsules containing many transversely tubercled seeds. Occurs as a weed throughout India.

The leaves are found to contain flavones such as apigenin, acacetin and luteolin, glycoflavones like methoxy orientin and its isomer, phenolic acids (vanillic, syringic, protocatechuic and gentisic acids), proanthocyanins and tannins. The seeds yield oil (18%), protein (14%), luteolin, β-sitosterol and cyanolipids.

The plant is a tonic and used in insomnia, inflammations, respiratory troubles, and as a diuretic and antiseptic. The whole plant is useful in chronic skin troubles. The leaves are said to the hypoglycemic and, therefore, used in diabetes. The roots are useful in fevers, gonorrhoea and lithiasis. The plant is eaten to induce sterility in man.

Woodfordia floribunda Salisb. (Lythraceae)

Syn. *W. fruticosa* Kurz.

Dhataki, Woodfordia

W. floribunda is a straggling leafy shrub with white pubescent young branches, opposite sessile, ovate-lanceolate velvety leaves having an intramarginal vein, deep orange-red flowers in 2-15 flowered cymes arising from the axils of leaf scars in leafless portions of stem and an ellipsoid capsule covered by the calyx enclosing many small obovoid seeds. Dhataki is common on rocky slopes, river banks, all over India, Africa and S.E. Asia.

Leaves contain tannins (12-20%) and lawsone. The bark contains 20-27% tannins. Flowers yield glycosides of cyanidin, pelargonidin, quercetin, myricetin and chrysophanol, ellagitannins and β-sitosterol.

Flowers of dhataki are used as a blood purifier and specific to uterine diseases like menorrhagia and leucorrhoea. It is also used in preventing abortion, healing ulcers, dysentery and diarrhoea. Leaves are also used as a tanning material.

Gardenia gummifera Linn. f. (Rubiaceae)

Dikamali, Gardenia

Dikamali is a tall shrub with resinous buds, elliptic-oblong, opposite sessile leaves, connate stipules, white large solitary flowers (changing to yellow) in axils and a beaked ribbed fruit with many small seeds. Dikamali is common in western India.

The resin and leaves contain flavones such as gardenin, dimethyl tangeretin, nevadensin, wogonin, isoscutellarien, apigenin and demethoxy sudachitin. Stem bark yield oleanolic aldehyde, sitosterol, erythrodiol and 19-OH erythrodiol.

Dikamali is a well-known antiseptic and carminative and is used in dyspepsia and as an anthelmintic.

Gardenia resinifera Roth.

Syn. *G. lucida* Roxb.

Dikamali

G. resinifera, another source of Dikamali, is a small tree resembling *G. gummifera*, but with petiolate leaves.

The resin contains flavones such as gardenins, A, B, C, D, & E, desmethyl tangeretin, nevadensin, acerosin, wogonin, demethoxy woganin, trimethoxy woganin, hydroxywoganin, apigenin, demethoxy sudachitin and 5,7,3',5'-tetrahydroxy-8,4'-dimethoxy flavone.

Bauhinia purpurea Linn. (Caesalpiniaceae)

Kachanar, Butterfly Tree

B. purpurea is a medium-sized evergreen tree with bifid coriaceous, ovate leaves shallowly cordate at the base, pink/white flowers in terminal panicles and long flat distinctly reticulate beaked pods containing 12-15 seeds. Cultivated throughout India for the flowers and beautiful foliage.

Leaves contain quercetin, quercitrin, rutin and apigenin. Flowers are found to contain astragalin, isoquercitrin, quercetin and glycosides of pelargonidin. Seeds yield appoximately 16% fat, 27% protein, 15% carbohydrates and 15% oil, with linoleic (49%), palmitic (18.5%), stearic (18%) and oleic (11%) acids), besides the chalcones, butein and 3,4 dehydorxy chalcone and DOPA (2%). A phytohemagglutinin, trypsin inhibitor and an alkaloid also are reported.

The stem bark in used extensively in glandular diseases and as an antidote to poison. It is also anthelmintic and cures ulcers, swellings, leprosy, menstrual disorders and prolapse of the rectum. Leaves are hypoglycemic. Flower buds, which are used as a vegetable and also pickled, are considered laxative and anthelmintic.

Bauhinia racemosa Lam.

Ashta

This yellow-flowered *Bauhinia* is small bushy tree with drooping branches. The leaves are broader than long and pods are falcate and glabrous.

The bark contains a tetracyclic 2,2-dimethyl chroman, de-O-methyl racemosal, β-amyrin, β sitosterol and octacosane. Seeds contain protein (10%), mucilage (4%) and oil (Prabhakar, et al. 1994).

The bark is highly astringent and is also antidysenteric, antiinflammatory and cholagogue. The leaves are anthelminitic and used for diarrhoea.

Bauhinia tomentosa Linn.

Kachanar, Buddhist Bauhinia

B. tomentosa is a medium-sized deciduous tree with a grey bark, broad leaves, variously coloured flowers in short-peduncled corymbs and 10-15 seeded long hard flat pods. Kachnar is a common garden tree found all over India.

The flowers contain flavonoids such as isoquercitrin (6%), rutin (4.6%) and traces of quercetin. Seeds yield proteins (19%), mucilage (7%) and saponins.

The fruit is used as a diuretic and seeds, a tonic. The root bark is found useful in liver troubles and as vermifuge. Leaves are used as dyes.

Bauhinia variegata Linn.

Kovidara, Mountain Ebony

This is also a medium-sized deciduous tree, with sub-coriaceous broad leaves, variously coloured flowers borne in short-pedunded corymbs and long hard glabrous flat pods containing 10-15 seeds. This medicinal tree is normally cultivated in gardens for its showy flowers.

The stem contains β-sitosterol, lupeol and glycosides of kaempferol and 5,7-dimethoxy flavonone. The bark also yields tannins (9.6%). Violet flowers are found to contain cyanidin, malvidin and peonidin while white flowers yield kaempferol. Seeds yield protein (44%), pentosan (11%), a fatty oil containing oleic (32%), linoleic (36%), palmitic (17%) and stearic (13%) acids. Mucilage and phytohemagglutinins also are reported from seeds.

The root is used in dyspepsia, flatulence and to prevent obesity. The bark is antidysentric, tonic and anthelminitc. Flowers and flower buds are used as vegetables and in the treatment of diarrhoea, dysentery, piles, tumours, heamaturia, and menorrhagia and as a laxative. Bark is used for dyeing and tanning.

Cardiospermum halicacabum Linn (Sapindaceae)

Indravalli, Ballon Vine

C. halicacabum is an annual scrambling climber with wiry stems, deltoid bipinnate leaves (leaflets deeply dentate or lobed), small white polygamous flowers borne in umbellate cymes. The lower pair of peduncles of the inflorescence is modified to tendrillar hooks and the fruit is an inflated pyriform 3-valved capsule containing 3 globose smooth seeds having a white heart-shaped aril. A native of tropical America, this plant runs wild in forests and fences.

The leaves are found to contain flavones such as apigenin, acacetin, 7-O Me apigenin, 7,4'-diOMe apigenin and 3', 4,-diOMe luteolin, phenolic acids such as vanillic, syringic, melitotic, *p*-coumaric and ferulic acids, β-sitosterol, saponins and alkaloids. A cyanolipid is also reported from this plant. The seeds yield a fatty oil (28%) unique in having 11-eicosenoic acid as the major fatty acid (42%) and oleic, linolenic, arachidic and linoleic acids as minor components.

The roots and leaves are diuretic, laxative, stomachic and alterative and are recommended for hair growth, rheumatism, nervous disorders, piles, fever, etc. The plant extract exhibited significant diuretic and anti inflammatory effects as well as a sedative effect on the central nervous system.

Canscora decussata Roem. & Schult. (Gentianaceae)

Shankhapushpi, Canscora

C. decussata is a slender erect annual with quadrangular winged stems, ovate-lanceolate leaves, white flowers in leafy cymes and oblong capsules containing brown seeds. It is common in damp localities in central India.

The whole plant/aerial parts contain xanthones, glycoflavones such as vitexin and isovitexin, terpenoids like (-)loliolide, gluanone, canscoradione, friedlin, β-amyrin, sitosterol, stigmasterol, carotenoids, phenolic acids, sugar alcohols (bornesitol) and alkaloid gentianine and another glycoalkaloid.

The plant is laxative, tonic and febrifuge. It is reported to promote memory power, cure leprosy, oedema and urinary disorders and is hypertensive.

Glycine max Merril (Fabaceae)

Syn. *G. soja* Maxim,
Soja hispida Moench.

Bhat, Soybean

Soybean, a very important food and forage crop of Asia, is an erect annual having trifoliolate leaves with large ovate-lanceolate hairy leaflets, blue-violet mostly inconspicuous flowers in short axillary racemes and broad pods containing 2-5 compressed, spheroidal/ellipsoidal yellow/brown/white seeds. A native of China, soybean is now widely cultivated.

The seeds contain carbohydrates (35%), protein (45-50%), oil (20%) saponins, sterols, tocopherols, phospholipids, lectins and isoflavonoids. The oil consists of linoleic (50%), oleic (35%), linolenic (6%), palmitic (6%) and stearic (4%) acids. Six kinds of group A saponins Aa, Ab, Ac, Ad, Ae and Af giving soyasapogenol A as aglycone and 5 group B saponins, Ba, Bb, Bc (all three giving soyasapogenol B as aglycone), Bd and Be (soyasapogenol E) are reported from the seeds. The steroids present are stigmasterol and β-sitosterol. Several isoflavone glycosides genistein, diadzin (glycitein 7-O-β-D-glucoside), glycitein, 6''-O-acetyl genistein, 6''-O-acetyl diadzin and their aglycones, and 6-O-malonylated derivatives of diadzin, glycitein and genistein also are found to be present. Genistein, diadzin and their derivatives form 0.25% of the defatted meal.

Soybean oil is known to decrease thrombosis and found useful in treating cholethiasis. Soymilk is useful in treating acute diarrhoea (in infants) and atherosclerosis. Isoflavones are estrogenic and, therefore, are useful in treating breast cancer. This is done through preventing the effect of protein tyrosine kinase (PTK) activity. Some other effects of genistein are inhibition of topoisomerase 2 and accumulation of protein associated single-stranded breaks of DNA in the entire cell. Genistein is also reported to prevent leukaemia and prostate cancer by arresting the cell cycle, including apoptosis, modifying cellular differentiation programme and by modulating cell-cycle events. Genistein-induced apoptosis of prostrate cancer is preceded by a specific decrease in total adhesion kinase activity. Use of genistein is found to prevent incidence of cancer of colon, lung, liver, oesophagus, oral epithelium and bone marrow.

Iris germanica Linn. (Iridaceae)

Pushkarmula

This is a perennial herb having a bulbous root stock, basal sword-shaped leaves in two rows, flowers in racemes, borne on peduncle (sepals coloured, stigma crested) and a 3/6-ribbed capsule containing numerous seeds. A native of Europe, Pushkarmula is cultivated in the Himalayan ranges.

Rhizomes contain isoflavones irisolone, irilone, tectoridin, homotectoridin, and a benzophenone, 2,4,6,4'-tetrahydroxy benzophenone, flavones and sterols like β-sitosterol and α-and β-amyrins. Petals contain embinin, a glycoflavone (Dhar and Kalla, 1974).

The roots are used for cough, dyspnoea, anaemia, cardiac disorders, skin diseases, dysmenorrhoea and dysurea. It is also an aphrodisiac, diuretic and febrifuge. This is one of the components of an oral contraceptive.

Pterocarpus marsupium Roxb. (Fabaceae)

Asanah, Indian/Malabar Kino

This is a large deciduous tree with yellowish grey bark, imparipinnate leaves having 5-7 elliptic pubescent, alternate leaflets with emarginated apex, yellow papilionaceous flowers having crisped margins in terminal panicles and a circular winged samaroid fruit (wing veined) containing a subreniform seed. *P. marsupium* is a common tree in the forests of India and nearby countries.

The bark of this tree contains *l*-epicatechin. Heartwood yields liquiritigenin and isoliquiritigenin. Other compounds isolated are a number of C-glycosides like pterocarposide, pteroisoauroside (isoaurone C-glucosides), marsuposide, benzofuranone C-glucoside, pteroside (a benzofuran C-gluoside), vijayosin [1,2-bis(2,4-dihydroxy,3-C-glucopyranosyl)ethanedione],a glucoside of 2,6-dihydroxybenzene and a sesquiterpene. Kino contain kinotannic acid, phlobaphenes, protocatechuic acid, resin and gallic acid (Handa et al. 2000; Maurya et al. 2004).

Heartwood is well known for its activity against diabetes, leprosy and other skin diseases. It is also useful in diarrhoea and haemophilic disorders. The bark and leaves are used in scabies, leucoderma, leprosy and skin diseases. The gum-resin "Malabar Kino" is useful in diarrhoea. Clinically, the wood is found to be an effective hypoglycaemic and hypocholesteremic.

Tephrosia purpurea Pers. (Fabaceae)

Sarapunkhah, Wild Indigo

This is a heavily branched undershrub with imparipinnate compound leaves (leaflets small, elliptic or oblanceolate, mucronate, clothed by silky hairs), purple-coloured papilionaceous flowers borne on axillary or terminal racemes, and a falcate compressed mucronate pod containing 5-6 ovoid seeds. *T. purpurea* is a common plant in waste places.

All the parts of the plant contain rotenoids. Leaves contain rutin, quercetin, 4'-OMe quercetin and phenolic acids such as ferulic , *p*-OH benzoic , vanillic, syringic and melilotic acids. Seeds contains galactomannans and rotenoids.

The roots or even the whole plant is used against enlargement and inflammation of liver and spleen. It is an useful anthelmintic, diuretic, laxative, diuretic and is effective against skin diseases. Roots are useful in hydrocele, diarrhoea and colic. Seeds are anthelmintic.

Derris indica Bennet (Fabaceae)

Syn. *Pongamia pinnata* Vent.
P. glabra Pierre

Karanj

Karanj is a tall tree with a soft bark, alternate imparipinnate compound leaves (leaflets usually 5, elliptic acuminate), pinkish-white flowers in axillary racemes and a woody compressed indehiscent pod narrowed at the base with a short decurved mucro at apex enclosing a single reniform seed. A common tree in western India, Karanj grows wild or cultivated.

The stem bark contains chromenoflavone, pongachromene, glabrin, glabra-I, glabra-II, karanjin, pongapin, kanugin and demethoxy kanugin. The wood also contains same compounds. The leaves yield a furanoflavone, 3'-OMe pongapin, karanjin, kanjone and other known flavonoids. The flowers contain another furanoflavone pongaglabol, aurantiamide-OAc and β-sitosterol. The fruits yield furanoflavonoids pongapinnol A-D and a coumestan, pongacoumestan. The seeds yield a fixed oil rich in oleic (70%) and linoleic acids and flavonoids such as dimethyl chromenoflavanone, isolonchocarpin, demethoxy kanugin, pongol, a β-diketone karanjin, pongapin, lanceolatin B, etc. (Yadav et al. 2004).

The seed oil is antiseptic and used in leucoderma. The seeds are used as expectorant and in whooping cough. A paste of the seeds is used in leprous sores and painful rheumatic joints. It is also used as a febrifuge and a tonic in asthmatic and debilating conditions. The roots are piscididal. The plant is used in leprosy and is an effective remedy for all types of skin diseases like scabies, eczema and ulcers.

Mallotus philippensis Don. (Euphorbiaceae)

Kamala, Kamila Dye Tree

Kamila dye consists of the fruit hairs of *M. philippensis* Don, a native of Asian and African tropics. It is a monoecious tree with ovate-elliptic leaves. Male flowers are in axillary racemes. Capsules and the seeds are covered with minute red stellate hairs and a red resinous powder.

The dye contains rottlerin (a complicated derivative of chalcone), isorottlerin, resins and waxes. The bark and wood contain about 0.5% bergenin (an isocoumarin), folic acid, α-amyrin, and sitosterol and acetycl alcur.

The dye is a bitter anthelmentic and cathartic.

Silybium marianum Gaertn. (Asteraceae)

Milk Thistle

This native of California is a glossy-leaved, erect, thistle-like annual herb with stem-clasping, large, lobed leaves which are white spotted above. The purplish flowers are arranged in large terminal solitary subglobose heads having a fleshy receptacle and spiny bracts. The fruit is an achene with shiny pappus.

The seeds contain three flavonolignans, silybin, silydianin and silychristin, collectively known as silymarin.

Silymarin is the only drug possessing genuine antihepatotoxic properties, i.e. a protective effect on the liver. It is antagonistic to α-ammanitine poisoning. The whole plant and seeds are used in jaundice and other biliary affections. It is also useful as a galactogogue, for fever, dropsy and uterine problems. Roots and leaves are used as pot herbs and salads. Diabetics consume the flowering heads. Seeds are used as a substitute for coffee.

4.12 Quinones

Quinones, the aromatic diketones, form the largest class of natural colouring matters. Of the total 500 and more compounds known, about 50% occur in higher plants and of the rest, a large percentage are known from fungi. Though quinones are the sole colouring pigments of fungi, in higher plants they play a subsidiary role, being mostly confined to the bark or underground portions and, if present in other parts, make very little contribution to the colour, due to the masking by other pigments. Colourless quinones and their derivatives also exist in plants possessing different physiological roles.

Quinones may be benzo, naphtho or anthraquinones, depending on the mono-, bi-or tri-cyclic ring systems they possess.

4.12a Benzoquinones

Benzoquinones possess an aromatic ring with or without an aliphatic side chain. A few of them are medicinally important, e.g. embelin, a dihydroxyquinone isolated from the dried fruits of *Embelia ribes* Burm. f., is an anthelmintic used against tapeworm and for skin diseases. The physiologically active representatives of this group are plastoquinones, ubiquinones and tocopherols. Plastoquinones are believed to act as redox carriers in photosynthesis; ubiquinones as hydrogen carriers in respiratory chain and tocopherols as antioxidants or growth regulators.

4.12b Naphthaquinones

Most of the naphthaquinones are yellow or red plant pigments. A few, like chimaphilin, plumbagin and eleutherin, are toxins or antimicrobial agents; while some others like lapachol are important dye stuffs. Vitamin K and related compounds are have specific roles in electron transport systems. Arnebin, another naphthaquinone isolated from the roots of *Arnebia nobilis* roots, is a potent anticancer agent.

4.12c Anthraquinones

Of this large group of quinones, about 50% are distributed in higher plants, especially Rubiaceae, Rhamnaceae and Polygonaceae, and the rest in fungi. This group provides some important dyestuffs and many purgatives.

Hydroxylated anthraquinones occur in the form of glycosides. Carboxylated anthraquinones also are reported. A reduced form of anthraquinone, anthrone and its dimer, dianthrone exist in many plants. Dianthrones may be homodianthrones and heterodianthrones, depending upon the type of monomers they possess. But the pharmacologically active anthraquinones are found to be anthranols that are formed as a result of slow hydrolysis of the glycosides during storage. Anthranols also, in course of time, get oxidized to anthraquinones. The pharmacological action of anthraquinone is due to its conversion to anthranols by intestinal bacteria (Fairburn, 1964).

In plants, their function is not properly understood. It is assumed that they play some role in oxidation-reduction reactions. For further details about these compounds, the reader is referred to Thomson (1971).

Embelia ribes Burm.f. (Myrsinaceae)

Vavding, Embelia

E. ribes is a large woody scandant shrub with long internodes, bark studded with lenticels and coriaceous, elliptic-lanceolate leaves (mature leaves covered with minute reddish sunken glands), small greenish-yellow flowers in lax paniculate racemes and small globose 1-seeded fruit tipped with persistent style. Embelia can be seen throughout India in forests.

The fruits, the drug, contain a benzoquinone, embelin (2,5-diOH-3-undecyl-1,4-benzoquinone).

Embelin and embelia are well-known anthelmintics, especially for tapeworm. The fruits are effective in cystic and abdominal tumours. With *Piper longum* and asafoetida it forms a well-known contraceptive. *Embelia* is a male contraceptive plant in the sense that embelin impairs spermatogenesis to the level of infertility (effect is reversible). Embelin exhibits significant anti-implantation and post-coital antifertility activity also in animals. With leaves of *Ficus religiosa* and borax in cow's milk, is a quick aborticide.

Arnebia nobilis Rachinger (Boraginaceae)

Ratanjot

A. nobilis is a herbaceous perennial with purple roots and cauline and radical rough leaves. The flowers are clustered in cymes and nutlets are ovoid. *A nobilis* is a native of the Middle East.

The roots, which are generally covered by several layers of a thin scaly bark, yield up to 2.5% of a red dye consisting of seven napthaquinones, arnebins 1-7. Wax, β-sitosterol and heptacosanoic acid are the other constituents of the root.

The roots are used for dyeing textiles such as wool and silk and also as a food colour. Arnebins are known to possess anticancer activity wherein crude extract/arnebin-1/arnebin-3 is found to inhibit Walker Carcino-Sarcoma tumour cells. They also exhibit antibacterial and antifungal activities. Ratanjot is used medicinally for liver inflammation and as an anthelmintic as well as an antiseptic to wounds, sores and burns.

Stereospermum suaveolens DC. (Bignoniaceae)

Patala

Patala is a large deciduous tree with pubescent branches, large opposite imparipinnate leaves having broadly elliptic acuminate rough leaflets, moderately large purple campanulate 2-lipped flowers (lower inside of corolla is bearded) and long cylindric, white-dotted capsules containing winged large seeds. This tree is common on the Western Ghats.

The bark yields a gum and all parts contain a napthaquinone, lepachol. The wood also contains lepachonone. Seeds yield a fatty oil and β-sitosterol.

All the parts are used in medicine. Roots form a component of *Dasamula* (ten roots), which is a general tonic. The roots are diuretic, cardiotonic, expectorant and effective in piles. Flowers are used in bleeding diseases, sore throat and diarrhoea. Fruits are useful in blood-diseases.

Stereospermum colais Mabb. is also used in place of patala.

Plumbago indica Linn. (Plumbaginaceae)

Syn. *P. rosea* Linn.
P. coccinea Boiss.

Chitrak, Plumbago

P. indica is an erect branched shrub with long tuberous roots, ovate-oblong short-petioled leaves, rose red flowers in long terminal spikes (calyx is glandular hairy outside) and a membraneous capsular fruit enclosed in the persistent calyx. This is a native of Sikkim, cultivated throughout for its medicinal purposes and as an ornamental.

The root bark contain a napthaquinone, plumbagin (0.91%), sitosterol, and tannins.

The roots are highly valued for their curative action on leucoderma and other skin diseases. They are digestive stimulants, diuretic, germicidal, vesicant and abortifacient. Chitrak is also used for haemorrhoidal inflammation of anus, diabetes, diarrhoea and elephantiasis.

Plumbago zeylanica Linn., a perennial herb with ovate leaves having an amplexicaul leaf base and white flowers is used as a substitute of chitrak.

Aloe barbadensis Mill. (Liliaceae)

Syn. *A. vera* Linn

Kumari, Indian Aloe

A. barbadensis is a succulent perennial with a short stem, fleshy narrow-lanceolate leaves having a dentate margin and yellow flowers in racemes borne on scapes.

The leaves contain mucilage, anthraquinones and chromones. The mucilage, on hydrolysis, yields glucose, galactose, mannose and galacturonic acid. The quinones present are barbaloin and isobarbaloin (C-glycosides) and their aglycone, aloe-emodin. Aloesone (7-hydroxy chromone) and its C-glycoside-aloesin are the chromones located. Also present are glycoproteins, alocutin A and B.

Fresh leaf juice is cathartic and also used for eye troubles, spleen and liver ailments, dermatitis and other skin diseases. The leaves are also credited with hypocholesteremic activity. Aloe finds use in menstrual and uterine disorders and stomach pain and as a tonic after pregnancy. Alocutin A and B are lectins found useful in cancer and inflammation.

Other species of *Aloe*, particularly *A. ferox* Mill., *A. africana* Mill. and *A. spicata* Linn. provide drugs similar to *A. barbadensis*. These plants yield lesser amounts of aloe-emodin but contain about 8% β-barbaloin.

Cassia Linn. (Caesalpiniaceae)

A number of species of *Cassia* are medicinally important. Many of them contain quinones or tannins or both and exhibit a wide variety of properties. The important species are the following.

Cassia occidentalis Linn. (Caesalpiniaceae)

Kasamarda, Coffee Senna

This is an erect undershrub reaching a height of 1.5 m with reddish stems, pinnate leaves having 3 pairs of glaucous membraneous lanceolate leaflets, yellow flowers in short racemes and recurved, compressed glabrous pods containing olive green, ovoid, compressed seeds. *C. occidentalis* is a common weed throughout India.

The leaves are found to contain anthraquinones such as chrysophanol, emodin (and their glycosides), physcion, metteucinol-7-rhamnoside, jaceidin-7-rhamnoside and 4,4',5,5'-tetrahydroxy–2,2'-dimethyl-1,1'-bianthraquinone. Also present are tetrahydroanthracene derivatives, germichrysone and occidentalins A & B. The pods yield glycosides of 7-OMe quercetin and 3,6,3'-trimethoxy quercetin. The seeds contain a gum (galactomannan), a fixed oil and anthraquinones such as 1,8-dihydroxy-2-methylanthraquinone, 1,4,5-trihydroxy-7-methoxy-3-methyl anthra-quinone, physcion, rhein, aloe-emodin, chrysophanol and steroidal glycosides based on campesterol and β-sitosterol. The roots yield bitetrahydroanthracene derivatives, occidentol-I and II, pinselin (cassiolin), rhein, aloe-emodin and their glycosides, chrysophanol, physcion, emodin, islandicin, helminthosporin, xanthorin, 1,8-dihydroxyanthraquinone, α-hydroxy anthraquinone, 1,8-dihydroxy-3-methoxy xanthone, quercetin, sitosterol, campesterol and stigmasterol. Also present in the roots are pinselin, questin, germichrysone, methyl germitorosone and singuenol.

All parts of this plant exhibit similar medicinal properties such as purgative, tonic, febrifugal, diuretic and also used for skin diseases. A bark infusion is given in case of diabetes. The entire plant is employed in dropsy and hepatic cirrhosis. Leaves are prescribed in cough, as an antiperiodic, vermifuge and anti-inflammatory. Seeds, used as a substitute of coffee, are blood tonics and useful in cough, heart ailments and cutaneous diseases. Root bark is used as a substitute for quinine to cure fever and a specific medicine for gonorrhoea and hepatic troubles. Germichrysone and occidentalins A & B are anticancerous in nature.

Cassia tora Linn.

Chakramarda, Sickle Senna

C. tora is a foetid annual shrub with pinnate leaves bearing 3 pairs of ovate-oblong leaflets (two pairs of leaflets have glands at their bases, all show sleeping movements), bright yellow flowers in pairs and stout long pods containing 25-30 green seeds. This is a very common weed throughout India.

The stem bark contains emodin, rhein, euphol, β-sitosterol and basseol. The leaves also contain emodin, kaempferol *D*-mannitol stachydrine and choline. The immature seeds yield brassinosteroids brassinolide, castasterone, typhasterol, teasterone and 28-norcastasterone and the monoglycerides, monopalmitin and mono-olein. The seed is rich in protein (24%), carbohydrates (49%), oil (5.4%) and contain glycosides of 1,8-dihydroxy-3-methyl anthraquinone, 8-OH,3-Me,1-OMe anthraquinone, two glycosides of 9,10-diOH,7-OMe,3-Me-1H naphthopyran-1-one, cassiamide, rhein, emodin, aloe emodin, physcion, rubrofusarin, nor-rubrofusarin, 8-hydroxy-3-methyl anthraquinone, chrysophanic acid, obtusin, toralactone, trigonelline, stachydrine and choline. Seed oil contain linoleic (45%), oleic (22%) and palmitic (22%) acids, 24-ethyl cholest-5-en-3β-ol, 24-ethyl cholest-5,22-dien-3β-ol and 24-ethyl cholest-7-en-3β-ol. The carbohydrates include a gum (galactomannan – 7.6 %) also. Roots yield choline, an anthraquinone, propelargonidin and β-sitosterol.

The whole plant is purgative, anthelmintic and exceedingly useful in eczema, ringworm and other skin diseases. Mature leaves, sometimes adulterated with senna, are anthelmintic and antiperiodic. Seeds are substituted for coffee beans, and are well-known tonics and stomachics. It is also used as a blood purifier.

Cassia senna Linn.

Sonamukhi, Alexadrian/Tinnevelly Senna

C. senna is an erect branching shrub of 1.5 m with pinnate leaves having lanceolate-elliptic leaflets in 3-9 pairs, brilliant yellow flowers in terminal racemes and dark brown flattened pubescent pods containing 5-7 obovate dark brown seeds.

The leaves and the shells of the fruits contain sennosides and related anthraquinones. Shells contain 3-5% of these compounds whereas the leaves have 2.5–4%. The various sennosides are sennoside A, B, C and G as well as three sennidins. Also present are free anthraquinones such as rhein, chrysophanol, emodin and aloe-emodin. Leaves contain monoglucosides of rhein and aloe-emodin whereas the pods contain monoglucosides of chrysophanol, rhein, aloe-emodin, and 6-hydroxymusizin. The flavonoids present are isorhamnetin and kaempferol as glycosides. Leaves also contain 7% mucilage. Roots are found to contain physcion, physcionin, chrysophanin and sennidin C.

The leaves and shells are used as tonics in Ayurvedic and Unani systems of medicine. This is an efficient purgative but due to the tendency to cause gripe, it is usually administered along with carminatives. Other uses of senna include as a febrifuge, in spleenic enlargements, anaemia, jaundice, rheumatism, tumours, and bronchitis and against skin diseases.

Cassia fistula Linn.

Amaltas, Indian Laburnum

C. fistula is a deciduous medium-sized tree with a grey smooth bark, large pinnately compound leaves bearing 4-8 pairs of stalked ovate leaflets having a silvery pubescence, bright yellow flowers in axillary lax pendulous racemes and pendulous cylindrical black pods (upto 60 cm length) containing hard brown biconcave seeds embedded in a sweetish pulp.

The bark contains flavonoids such as 3, 5, 7, 3',4'-pentahydroxy, 6, 8-dimethoxy flavonol, 3, 5, 7, 4'-tetrahydroxy,6, 8, 3'-trimethoxy flavonol, 1, 3, 8-trihydroxy,3, 7-dimethoxy xanthone, (+)-catechin, epicatechin, fistacacidin, fistucacidin, kaempferol, procyanidin, propelargonidin, rhein glycoside and steroids like lupeol and β-sitosterol. The leaves yield sennosides A and B, chrysophanol, physcion, rhein, kaempferol, quercetin, epiafzelchin and epicatechin and their polymers; gallic, ellagic and protcatechuic acids. The roots are found to contain 7-methyl physcion, oxyanthraquinones, betulinic acid, β-sitosterol, fistucacidin, tannins, and phlobaphenes. Flowers contain kaempferol, propelargonidin tetramer, rhein, fistulin rhamnoside, stigmasterol, 28-isofucosterol and methyleugenol. The pulp from the pods is rich in protein (upto 20%), and carbohydrates (26%), and contains sennosides A and B, rhein and its glucoside barbaloin, aloin, pectin, tannin and sugars. Pods yield procyanin B-2, epicatechin dimers, epiafzelchin dimers and epicatechin-epiafzelchin dimers. The seeds yield a gum and a fixed oil containing cyclopropenoid fatty acids such as vernolic, malvalic and sterculic acids (Gupta et al. 1989; Asseleiah et al. 1990).

The bark is used as a tonic and against dysentery and various skin diseases. It is also useful in leprosy, jaundice, syphilis and heart diseases. The pods are a well-known laxative. The shell showed anti-fertility activity in mice and is found to cause abortion and expulsion of the placenta. The pulp from the fruits is a safe purgative for children and pregnant women. It is also an analgesic, antipyretic (against malaria) and used against anthrax, dysentery, leprosy and diabetes. Seeds and leaves are used in preparation of a drink, which is carminative, laxative and antipyretic. The roots are useful in rheumatism, hemorrhages, ulcers, skin diseases and leprosy. The flowers are said to the purgative, febrifugal and anti-bilious.

Cassia angustifolia Vahl (Caesalpiniaceae)

Hindisana, Senna

C. angustifolia Vahl, a native of Arabia and India, constitutes senna. The former plant is a subglabrous erect shrub with stipulate pinnate leaves having 8-12 obvate-oblong leaflets, pale yellow flowers in narrow peduncled cymes and a flat glabrous recurved pod containing 6-12 seeds.

The leaflets yield about 2% quinones consisting of aloe-emodin, rhein, sennosides A, B, C & D and many other heterodianthrones. Also present in the drug are flavonols such as kaempferol, sterols, mucilage and resins.

Senna is widely used as a purgative. It is preferred over rhubarb because it does not produce the astringent after effect of the latter.

Senna pods. The fruits of *C. angustifolia* and, *C. acutifolia* Delile are used as purgatives. The active principles are very similar to those of the leaves.

Rheum palmatum Linn. (Polygonaceae)

Rhubarb

Rhubarb is the dried rhizome, deprived of bark, of *R. palmatum* and many other species of *Rheum*. Rhubarb is a stout perennial herb with clumps of large broadly cordate and deeply palmately lobed radical leaves and green, short pedicelled flowers in leafy panicles borne on flowering shoots. These plants, natives of Hong Kong, are widely cultivated in China.

The drug yields a variety of anthraquinones, which occur in the form of glycosides or aglycones. They are: (1) dihydroxy phenols such as chrysophanol; trihydroxy phenols like emodin, aloe emodin and methoxylated phenols like physcion and their glycosides; (2) Carboxylated anthraquinones, e.g. rhein (cassic acid) and glucorhein; (3) Anthrones and dianthrones based on chrysophanic acid, rhein or physion. Sennidin is a dianthrone glucoside; and (4) Heterodianthrones derived from 2 different anthrones like palmidin A (aloe-emodin anthrone and emodin anthrone), palmidin B (aloe-emodin anthrone and chrysophenol anthrone), etc.

Rhubarb contains a large amount of both hydrolyzable and condensed tannins, as well as starch and calcium oxalate.

The drug is used as a bitter stomachic and in the treatment of diarrhoea because of the purgative action of anthrones and the astringency of tannins.

R. emodi Wall., *R. webbianum* Royle, or other species of *Rheum* native to India are known in commerce as Indian Rhubarb. They contain upto 2.2% anthraquinones. Emodin and chrysophanol, along with their derivatives, form the major constituents. Also present are rhein, aloe-emodin, physcion and their glycosides as also hydrolyzable and condensed tannins. The principles are similar to *R. palmatum*. This drug is a purgative, useful in treating diarrhoea and shows anti-inflammatory, diuretic and anti-cancer properties.

Rhamnus purshiana DC. (Rhamnaceae)

Cascara Bark

Cascara bark is derived from a deciduous tree native of the USA and Canada. The leaves of this plant are elliptic or oblong, tufted at the tips of the branches and flowers are borne in pubescent umbels. The fruits are black drupes.

For better preparations of the drug, the bark is to be stored for at least one year, during which period a slow hydrolysis of the constituents is believed to take place. The drug contains about 6-9% anthraquinones, mostly as glycosides. The C-glycosides amounts to 70-80% of the total glycosides and O-glycosides 10-20%. The different anthraquinones present are: (1) Cascarosides—The primary glycosides: cascarosides A and B are the most important compounds in this drug. They are the O-glycosides of barbaloin—a C-glycoside. Cascarosides C & D are also active compounds and are O-glycosides of chrysaloin. (2) Barbaloin and chrysaloin present in the free form. (3) Emodin, emodin oxanthrone, aloe emodin, chrysophenol, etc., free or as O-glycosides and (4) Homodianthrones of emodin, aloe-emodin and chrysophenol and heterodianthrones like palmidin A, B and C.

Cascara is a purgative having an action very similar to that of senna. Aloe-emodin isolated from the seeds of *R. frangula* Linn. exhibited anticancerous property in the sense that it showed significant inhibitory activity when tested in mice against the P-388 lymphocytic leukaemia.

Rhamnus alnus Mill (Rhamnaceae)

Alder Buckthorn Bark, Frangula Bark

The bark of Alder buckthorn, a native of Europe and West Asia, provides the drug. The plant is a deciduous tree attaining a height of 5 m with broadly ovate shining leaves, greenish yellow flowers in sessile glabrous umbels and reddish black drupes.

Glycosides form the major group of the 2-5% anthraquinones present in the purple bark. The principal compounds in the drug are a pair of isomers frangulosides A & B, together known as frangulin or rhamnoside-franguloside. The minor compounds are frangulin B, emodin dianthrone and its glycoside, and palmidin C and its rhamnoside. The seeds contain aloe-emodin.

Alder buckthorn bark is used as a cathartic.

R. cathartica L. and *R. carnifolia* B & H. also possess active quinones and so are used as cathartics.

Andira araroba Aguiar (Fabaceae)

Chrysarobin, Goa Powder

The wood of *A. araroba*, a native of Brazil, but extensively cultivated in Goa, provides chrysarobin. This plant is a large tree with glabrous branches; odd pinnate leaves having oval leaflets, white flowers in racemes and an oblong pod.

The yellowish brown powder found in large lacunae of the stem wood forms the drug. This powder, which is scraped off, contains 50-75% chrysarobin, 7% bitter substances and about 2% resin. Chrysarobin is an anthrone-mixture consisting of 40% chrysophenol-anthrone, 20% monomethyl ether of emodin-anthrone and about 30% monomethyldehydroemdinan-throne.

Even though chrysarobin possessses a laxative action, it is seldom used thus. The drug is extensively used in fighting skin diseases.

Rubia cordifolia Linn. (Rubiaceae)

Mangishta/Indian Madder

R. cordifolia is a scabrid hispid perennial climbing herb with 4-angled rough stem, very long cylindric red roots, numerous scandant divaricate or deflexed branchlets, ovate acute scabrid leaves in whorls of four (one pair of each whorl larger, as also the lower leaves are large, margins with minute white prickles, petioles also with sharp recurved prickles), greenish small flowers in axillary or terminal cymes and small globose purple fruits containing two seeds. Indian madder is common in hills throughout India, Africa, Malaysia and Japan.

The drug contains two anthraquinones, purpurin (trihydroxy anthraquinone) and munjistin (xanthopurpurin–2-carboxylic acid) in major amounts and xanthopurpurin (purpuroxanthin) and pseudopurpurin (purpurin-3-carboxylic acid).

Mangishta is one of the acclaimed blood purifiers and used as an alterative, tonic and for diseases of blood, skin and urino-genital system. It improves complexion, and the microcirculation within the body and cures jaundice, paralysis, freckles, etc.

Pterocarpus santalinus Linn. (Fabaceae)

Raktachandan, Red Sandalwood

P. santalinus is a tall tree with 3-5 foliolate imparipinnate leaves (leaflets large, ovate-orbicular, apex emarginated and alternate on rachis), small papilionaceous flowers in axillary racemes and a broad orbicular winged pod containing a single subreniform seed. This tree is common in the Western Ghats.

The wood of this tree contains a quinone, santalin (santalic acid 16%), desoxy santalin (a napthaquinone), santal (isoflavone), two aurone glycosides (6-OH, 5-Me, 3',4',5'-triOMe aurone 4-O-α-L-rhamnopyroside and 6,4'-diOH aurone-4-O-rutinoside) and colourless pterostilbenes, pterocaspin and homopterocarpin (Kesari et al. 2004).

Raktachandan is a reputed remedy to remove pimples, scars and other skin diseases. It purifies the blood and is an antipyretic, diaphoretic and febrifuge. It is also useful in haemophilic disorders and inflammations.

Hypericum perforatum Linn. (Hypericaceae)

St. Johns' Wort

H. perforatum is a perennial herb with erect 2-edged stem having slender stolons, glandular dotted oblong leaves, yellow flowers in trichotomously branched corymbose cymes (petals with black-glandular edges) and egg-shaped capsules containing numerous seeds. A native of Europe, it is also seen in Africa and N.W.Asia.

The aerial parts of this plant contain hyperin, rutin and quercetin. Hypericin (hexahydroxy, dimethyl-napthadianthrone), the aglycone of hyperin, is also present in the flowers.

The herb is a spasmolytic, diuretic and stimulates gastrointestinal secretions especially bile. It has a very pronounced cicatrizing action on wounds, cuts and bruises. It is also used for body pain, in ulcers and abscesses, goiter and treatment of bed sores. Other uses include rheumatism, *anguina pectoris*, cardiac diseases, phlebitis, psoriasis, etc. It is also used in aerosol cosmetics useful for development and firming of breasts.

4.13. Tannins

Tannins are polyphenols having an astringent taste and the ability to convert putresible animal hides to hard stable leather. Two groups of tannins, hydrolyzable and condensed, are recognized in plants. The former group is soluble in water, contains simple phenolic acids esterified with one or more sugar molecules and are hydrolyzed by dilute acids to component compounds. Condensed tannins are polyphenols insoluble in water, which on treatment with acids, yield complex products of unknown composition called "tannin reds" or "phlobaphenes".

The common phenolic components of hydrolyzable tannins are gallic, digallic, and ellagic acids. The less common phenolics are hexahydroxydiphenic acid and chebulic acid. The sugar component is invariably glucose. The tannin molecules are often complex mixtures containing several different phenolic acids esterified to different positions of the sugar molecule and among themselves. Based on the phenolic nuclei involved, hydrolyzable tannins are further subdivided to gallotannins and ellagitannins. Free gallic acid and the various galloyl esters of glucose constitute the 'tannic acid', of commerce. Chinese gallotannin, obtained from the aphid galls of *Rhus semialata* Murr., contains a basic structure of 1,2,3,4,6-pentagalloyl glucose with 3-5 additional galloyl groups attached by depside linkages to form chains of 2-3 *m*-galloyl groups (Robinson, 1975). Tannins from other sources differ in the type and number of phenolic acids attached to the glucose molecule. Thus, tara tannin (an acidic tannin from the seeds of *Caesalpinia spinosa* Ktze) is a penta- or hexagalloyl derivative of quinic acid containing an *m*-trigalloyl group.

Acer tannin is 3,6-di-0-galloyl-1, 5-anhydro D-sorbitol. Hamamelitannin is another gallotannin of medicinal importance obtained from *Hamamelis.*

Corilagin is the simplest ellagitannin, which occurs in divi-divi (*Caesalpinia coriaria* Willd) and myrobalanns (*Terminalia chebula* Retz). Other ellagitannins contain, alongwith gallic acid and ellagic acid, compounds resembling hexahydroxy diphenic acid, e.g. Chestnut tannin (*Castanea* sp) and valonia (*Quercus macrolepis* Klotschy) contain dehydrodigallic acid and valoneaic acid, respectively.

The condensed tannins are a complex group of compounds, the components of which are extremely difficult to isolate. Till date, no complete structure of condensed tannin is known with certainty. Flavolans are undoubtedly the principal components of this group. There also exist flavonols, chalcones and flavonones, though to a limited amount. The flavolans—proanthocyanidins—yield anthocyanidins and phlobaphenes on acid hydrolysis, of which the latter is of a highly complex structure. It is assumed that a molecule of catechin is linked with flavan-3, 4-diol, in a head to head, tail to tail or head to tail fashion, by –C-O-or C-C-linkage to produce a molecule of proanthocyanidin. In addition to flavolans, hydroxystilbenes (e.g. picetatannol) have the ability to tan leather and are grouped among these compounds.

The hydrolyzable tannins are abundant in the leaves while the condensed tannins are concentrated in the wood. Usually a single species contains only one of these groups but there are instances where both types of tannins co-occur in the same plant. Thus, in *Schinopsis* sp., hydrolyzable tannins are predominant in leaves, whereas flavolans are present in heartwood, and the sapwood contains both.

Apart from the seeds, leaves, bark and wood, commercial tannin sources include roots (*Krameria triandra* R. & P.), tubers (*Dioscorea atropurpurea* Roxb.) and rhizomes (*Rumex hymenosepalus* Torr.). Many of the economically important products like areca nut and cocoa seeds also contain appreciable amounts of tannins. The astringency caused by many unripe fruits is due to tannins.

Tannins, being fungicidal, contribute much to the resistance of the plants against the attack of fungi and other microorganisms.

Tannin is used principally for the manufacture of leather. It is also used as a deflocculent to control the viscosity and gel strength of mud in oil-well drilling; in the manufacture of gallotannate ink, for floating certain ores, as antioxidants in edible substances especially fats, medical astringents and adhesive for plywoods and particle boards.

Barringtonia acutangula Gaertn. (Lecythidaceae)

Samudraphal, Indian Oak

B. acutangula is an evergreen tree with obovate-elliptic leaves having denticulate crenate margins and blade narrowing to petiole, fragrant flowers with many bright red stamens and quadrangular fibrous fruit containing a single large seed. A native of tropical Africa and Asia, Indian Oak is common in sub-Himalayan tracts and in Madhya Pradesh and Maharashtra.

The bark contains ellagitannins (16%), and saponins, and the leaves contain barringtonic acid, stigmasterol, β-sitosterol, β-amyrin, oleanolic acid, tangulic acid and acutangulic acid (Anjanayelu et al. 1978). Heartwood yield tanginol, barringtonic acid, dimethyl barringtogenate, barringtogenol E, barringtogenol C monobenzoate and barrinic acid (Barua et al. 1972). The fruit contains saponins based on barringtogenol B, C and D, methyl barringtogenate and methyl acutagenate (Barua et al. 1976). Seeds also contain barringtogenin and barringtogentin.

Leaves are used as a cure for diarrhoea. Fruits are anthelmintic, tonic and are used in gingivitis. The seeds are expectorant and emetic. The bark finds use in diarrhoea and blennorrhoea and as a febrifuge. The wood is useful as haemostatic in menorrhagia.

Barringtonia asiatica Kurz. (*B. speciosa* J.R. & G. Forst.), found in the Andamans, also contains similar compounds. The fruits contain anhydrobartogenic, 19-*epi*-bartogenic and bartogenic acids whereas the seeds yield a fatty oil (14%), hydrocyanic acid, barringtonin, barringtogenetin and A_1 barrinin. The fruits are narcotic and used as vermifuges.

Barringtonia racemosa Speng. another related plant seen in Konkan coasts, contains in its seeds, saponins (3%) based on barringtogenol, barringtogenic acid, R_1-barringenol and R_2-barringenol. The fruits are used for cough, asthma and diarrhoea; seeds are used in colic, ophthalima and jaundice. Bark, used as a vermifuge, contains 18% ellagitannins.

Hamamelis virginiana Linn. (Hamamelidaceae)

Witch Hazel

A native of Canada and USA, witch hazel is a spreading deciduous tree with elliptic-obovate leaves, bright yellow few-flowered racemes and dehiscent woody two-seeded capsules.

The leaves contain 'hamamelitannins' (a group of gallotannins), ellagitannins, gallic acid and traces of a volatile oil. Of the three hamamelitannins isolated, i.e. α-, β-and γ-; β-hamamelitannin is the most important one and is formed from 2 gallic acid molecules and one moelcule of the sugar, hamamelose. The volatile oil consists of 2-hexen-l-al, α-ionone and safrole.

The leaves are used as astringents and haemostatic due to the presence of tannins. Hamamelis water or witch hazel extract, prepared by the steam distillation of the leaves and young stems, is widely used for sprains, bruises, wounds and in eye lotions.

Syzygium cumini Skeels (Myrtaceae)

Syn. *S. jambolanum* DC.
Eugenia jambolana Lamk.

Jamun/Jambuh

S. cumini is a large tree with an exfoliating bark, elliptic-oblong leaves (with intramarginal vein), dull white tetramerous flowers crowded in panicled cymes from branches below the leaves (stamens many, ovary inferior), and a dark purple elliptic berry containing a globose seed. Jamun is common as a cultivated avenue tree or in forests of India, S.E. Asia and Australia.

The edible pulp from the fruits contains sugars, gallic acid, malic acid, tannins, glucosides of petunidin and malvidin and oleanolic acid. Seeds contain protein (8.5%), tannins (19%), ellagic and gallic acids, and a glycoside, jamboline. Flowers yield oleanolic acid and eugenia-triterpenoid A&B along with isoquercitrin, quercetin, kaempferol and myricetin. The stem bark contains betulinic acid, β-sitosterol, friedlin, tannins, myricetin, gallic acid and ellagic acid.

The fruit, especially the seeds, are extensively used against diabetes. The bark is used in dysentery and also used in cough, asthma, dyspepsia and hemorrhages. Seed extract is also used as a male contraceptive (Sinha et al. 1986).

Saraca asoca De Willde (Caesalpiniaceae)

Syn. *Saraca indica* Baker

Asoka

Asoka is a medium-sized evergreen tree with imparipinnate leaves having intrapetiolar stipule which sometimes close around the terminal bud. Leaves are copper red, pendulous when young, green later. Leaflets are in 4-6 pairs, oblong and lanceolate. Flowers are orange-red in dense axillary corymbose cymes. (Flowers are with two reddish bracteoles resembling calyx, an orange red calyx tube which is solid at the base, petals absent, ovary placed at the tip of the calyx tube on one side). Pods are black and woody, containing 4-8 ellipsoid seeds. A native of S.E. Asia, Asoka is commonly found cultivated in south India.

The bark yields tannins (6%), catechol, volatile oil, haematoxylin, a ketosterol, and saponins.

Asoka bark is widely used for treating uterine disorders such as haemoarrhages, dysmenorrhoea and menorrhagia. This is also used to cure inflammation and enlargement of cervical glands. Flowers are used for treating piles, scabies and skin diseases.

Thespesia populnea Sol. (Malvaceae)

Parisa/Palash, Portia Tree

T. populnea is a medium-sized tree with broadly ovate, cordate, acuminate, long-petioled leaves, solitary large yellow/orange/red (with a purple base) long-peduncled flowers having 3-5 lanceolate caducous epicalyx segments and globose, tomentose indehiscent capsules containing many ovoid, grooved, pubescent seeds. Common along the seashores, it can be seen planted as an avenue tree.

Bark contain tannins (7%), gossypol and a colouring matter. Leaves are found to possess acacetin ,3',4'-diOMe quercetin, and vanillic, syringic, gentisic, melilotic and ferulic acids (Sheeja, 1991). Flowers yield populnin and populnetin (both glucosides of kaempferol), herbacetin, quercetin, gossypetin, gossypol and β-sitosterol. Fruits also contain gossypol, and herbacetin. Seeds yield a fatty oil rich in linoleic and oleic acids, ceryl alcohol and β-sitosterol.

Bark is used mainly for treating skin diseases and urinary disorders. It is also used for piles, haemorrhoids and ulcers. Fruit, leaves and root also are used for psoriasis, scabies and other skin diseases. Leaf extract in oil is a remedy for inflammatory swellings.

Terminalia chebula Retz. (Combretaceae)

Hirda/Haritaki, Chebulic Myrobalan

This is a moderate-sized, much-branched tree with opposite elliptic-oblong large leaves (pubescent when young, have two glands at the top of the petiole), apetalous cream coloured flowers in axillary spikes and a glabrous obovoid, faintly five-ribbed fruit containing a bony stone. *T. chebula* is abundant in the Western Ghats.

The fruits contain 30-40% of tannins mostly as di-O-galloyl and tri-O-galloyl glucose. Some of the component tannins isolated are chebulagic acid ($C_{41}H_{30}O_{27}.10H_2O$), chebulinic acid ($C_{41}H_{30}O_{27}$) and corilagin. Also present are the breakdown products (partial) such as chebulic acid, 3, 6-digalloyl glucose, gallic acid, ellagic acid, terchebin (1, 3, 6-trigalloyl glucose) and 1, 2, 3, 4, 6-pentagalloyl glucose. Also present in the fruit are sugars (glucose and sorbitol), amino acids, and organic acids (succinic, quinic and shikimic acids).

Haritaki (the fruit rind) is a component of **triphala** (a group of three fruits—*T. chebula, T. bellerica* and *Emblica officinalis),* which is used as a *rasayana* (rejuvenative) drug. Haritaki is antiseptic, laxative, diuretic and carminative. This is one of the best laxatives, and used for diabetes, fever, anaemia, cardiac disorders, diarrhoea and dysentery.

Haritaki is one of the chief sources of tannin in India.

Terminalia bellirica Roxb. (Combretaceae)

Vibhitaki/Behada, Belliric Myrobalan

T. bellerica is a large deciduous tree with opposite broadly elliptic leaves crowded at the end of branches. Flowers, both unisexual and bisexual, are apetalous, yellowish green, borne in axillary spikes (male flowers towards the top). Fruit is a globose drupe, tomentose containing a single seed. This tree is seen throughout India.

Fruits are found to contain tannins, β-sitosterol, ellagic acid, gallic acid, chebulagic acid, ethyl gallate and sugars. Tannins present are both hydrolyzable and of condensed type. Seeds yield a fatty oil and protein.

The rind of the plant is one of the *triphala*, a general tonic. It is used for anaemia, cough, fever and is laxative and germicidal. It is considered to be useful in treating leprosy, tuberculosis, diarrhoea, inflammations and problems of eye, nose and throat.

This is also a source of tannin and dye.

Terminalia arjuna W.& A.

Arjun/Arjun Sadad

This is a large tree with a thick trunk, horizontally spreading branches and greenish white bark flaking off in large flat pieces. Leaves are large, subopposite, oblong, crenate-serrate and have 1-3 glands at the top of petiole. Flowers are apetalous, yellowish green in axillary spikes. Fruit is a five-winged drupe with veins of wings curved upwards. Arjun is common in Western India.

The bark is found to contain tannins (both hydrolysable and condensed, 20-40%), arjunic acid, a trihydroxy triterpene monocarboxylic acid, ($C_{39}H_{48}O_5$) and its glucoside arjunetin, β-sitosterol, friedelin and ellagic acid. Also present in the bark are CaO, MgO and Al_2O_3. Fruits contain 7-20% of tannin, arjunic acid, β-sitosterol, mannitol and KCl.

The bark of Arjun tree is a well-known cardiotonic. This is used in all sorts of cardiac failure and dropsy. It is also a good diuretic, febrifuge, tonic and antidysenteric.

Symplocos cochinchinensis Moore (Symplocaceae)

Lodhra, Symplocos

Lodhra is a medium-sized tree having elliptic-obovate, crenate leaves, small white fragrant flowers (corolla gamopetalous, stamens 12 or more, ovary inferior) in tomentose spikes and a small globose drupe having an apical ring, enclosing oblong seeds. A native of Southeast Asia, *S. cochinchinensis* is common in forests.

The bark is an acclaimed remedy for menstrual disorders, uterine problems and urinogenital diseases. It is a specific uterotonic and is useful in menorrhagia, leucorrhoea, diarrhoea, dysentery, fever, cough and gum diseases.

S.paniculata Miq (Syn. *S. crataegoides* Buch.-Ham.) is also used as lodhra in north India.

Adenanthera pavonina Linn. (Mimosaceae)

Kundhandana, Coralwood

This is a tall tree with bipinnate leaves having 4-8 elliptic-oblong alternate leaflets, yellow fragrant flowers in racemes and long pods enclosing scarlet biconvex seeds. Coralwood is quite common in India.

Seeds contain a fat (25-30%), protein (30%), stigmasterol, dulcitol and non-protein amino acids such a γ-methylene glutamic acid, γ-methylene glutamine and γ-ethylidine glutamic acid. The wood contains 2, 4-dihydroxy benzoic acid, robinetin, chalcone, butein and dihydroxyricetin. Leaves yield octacosanol, dulcitol, β-sitosterol and stigmasterol. Bark contains saponins based on oleanolic and echinocystic acids.

Seeds are used for treatment of cholera and paralysis. Seed extract is useful in pulmonary affections and chronic opthalmia. A decoction of leaves/bark is used against rheumatism and gout. Root extract is hypotensive. The wood yields a dye, which is used as a substitute for red sandalwood.

Acacia catechu Willd. (Mimosaceae)

Khair, Cutch Tree, Catechu

A. catechu is a small tree of 3m, with exfoliating dark grey bark, deciduous bipinnate leaves having a pair of recurved prickles as stipules, yellow flowers in long cylindrical spikes and glabrous pods. This is a common tree inhabiting the drier regions of India.

Catechu is a resin extracted by boiling the red heartwood in water. Kattha (pale catechu) is prepared by keeping the concentrated extract for a few days when the catechin settles (crystallizes) at the bottom. These crystals collected are moulded to various shapes and are called the kattha. The mother liquor, on concentration, yields Cutch (dark catechu). Kattha is almost pure catechin whereas cutch contains upto 60% tannins and 12-18% catechin. Kheersal is the crystalline or powdery deposits of catechin found naturally in older trees. Khair gum is naturally exuding and occurs in the form of pale yellow tears. It is sweet in taste, dissolves in water to form a strong mucilage and is a polysaccharide consisting of galactose, arabinose, rhamnose and aldobiurmic acid. Leaves contain upto 12% crude protein. Seeds possess protein (44%) and other extractives (40%).

Kattha is known for its antileprotic activity and is used as an astringent and in cases of cough and diarrhoea. It is extensively used in *pan* preparations. Kattha is a highly efficient antioxidant. Kheersal is used for the treatment of cough. Gum is a good substitute of gum Arabic.

Acacia nilotica subsp. *indica* Brenan (Mimosaseae)

Syn. *A. arabica* ***Willd.*** var. *indica* Benth.

Babul, Indian Gum Arabic

This is a moderate-sized evergreen tree, with dark brown deeply cracked bark; bipinnate leaves having spiny stipules, yellow fragrant flowers in globose heads and moniliform pods. Babul is common throughout the drier parts of India.

Babul bark is an important tanstuff in northern India. The tannin content of bark varies from 12-20% and consists of catechin, epicatechin, dicatechin, quercetin, gallic acid and procyanidin. The pods contain 12-19% tannin consisting of gallic acid, *m*-digallic acid, catechin, chlorogenic acid, galloylated flavan-3, 4-diol and robidandiol. The gum exuding from the wounds of the bark is similar to khair gum. Seeds contain protein (26%), other extractives 62%, and oil, 6%.

The bark of babul is used in asthma, bronchitis, leucorrhoea and pneumonia and is hypoglycemic. Gum stops bleeding and used for urinary and vaginal discharges. Pods are used in impotency and in urinogenital disorders. Various parts of *A. nilotica* are used in earache, cholera, dysentery, and leprosy and against worms.

Albizzia lebbeck Willd. (Mimosaceae)

Siris Tree

A. lebbeck is a large erect deciduous tree with dark brown bark, bipinnate leaves having 8-18 leaflets, greenish yellow flowers in globose heads and yellowish brown strap-shaped fruits containing 6-10 seeds. Siris is planted in many parts of India as a social forestry tree yielding timber.

The heartwood contains polyphenols such as melanoxetin, okanin, pro pelargonidin, lebbecacidin, melacacidin, pinitol and a saponin, lebbekanin E, based on a sapogenin acacic acid. Bark yields condensed tannins (7-11%) consisting of D-catechin, procyanidin, melacacidin and lebbecacidin, a saponin lebbekanin D (sapogenin, acacic acid) and steroids friedelin and β-sitosterol. The bark also yields a reddish brown gum. Leaves contain alkaloids, kaempferol, and quercetin and caffeic acid. Flowers yield a volatile oil consisting of *p*-nitrobenzoate, benzyl alcohol and benzoic acid and the saponin, lebbekanin D. Also present in the flowers are lupeol, α-and β-amyrins and a coloring pigment similar to crocetin. Seeds yield protein (40%), and saponins lebbekanin A & B based on sapogenins echinocystic and oleanolic acids.

A. lebbeck is said to have anti-dysenteric and anti-tubercular properties. The bark is used for bronchitis, leprosy, paralysis and worms and as an expectorant. Both bark and seeds are used as tonic and restorative and in diarrhoea and piles. The gum from the bark is used as a substitute for gum Arabic. Seeds and pods exhibit hypoglycemic and anticancerous properties also.

Albizzia procera Benth.

Safed Siris

A. procera is another large tree with pinnate leaves having 4-12 leaflets and a few flowered small heads in terminal panicles. Pods are thin, containing 8-10 seeds.

Roots are found to contain a saponin, based on oleanolic acid methyl ester, and α-spinasterol. The root bark yields gum, tannins (12-17%) and β-sitosterol. α-Spinaterol and oleanolic acid are the compounds present in roots while the leaves contain the former compound only. Stem bark contained β-sitosterol and isoflovones such as biochanin A, formonetin, genistein and diadzein. Stem wood also contains all these isoflavones besides a pterocarpan. Seeds are found to possess two saponins, proceranin (based on the aglycone machaerinic acid) and proceranin A (based on proceragenin A and B). An oleanolic acid saponin also in reported from seeds.

All the parts of the plant exhibit anti-cancer properties. The plant is used for intestinal diseases. The root saponin is found to possess spermicidal activity. A decoction of bark is used against rheumatism and haemorrhage. The gum obtained from this plant is similar to that of gum Arabic. Seed extract is hypotensive in nature.

Albizzia odoratissima Benth.

Kala Siris

This tree also yields a gum in the form of rounded tears. The stem and leaves contain tannin and seeds possess a saponin odoratissimin based on echinocystic acid and machaerinic acid. The bark is considered to be a tonic and effective in leprosy and ulcers. Leaves are used as expectorants and the bark yields a brown dye.

Butea monsperma Taub. (Fabaceae)

Syn. *B. frondosa* Konig.

Palas, Bengal Kino Tree

Palas is a deciduous tree with a crooked trunk, bluish-grey bark, long-petioled trifoliolate leaves (leaflets large, broadly obovate from a deltoid base) which are densely silky below but glabrous above, bright orange-red somewhat sickle-shaped flowers borne on long racemes on bare branches and silky tomentose pendulous pods enclosing a single flat reniform seed at the apex. A native of Indo-Malaya, Palas can be seen in most of the forests. It is also known as 'Flame of the forest'.

A red gum exudes from the natural cracks of bark and solidifies on exposure into elongated tears; butea gum or Bengal kino. It contains catechin, procyanidin, gallic acid and mucilage. The bark yield tannins (6%), 3-OH,9-

OMe pterocarpan (medicarpin), isoflavones such as 5-methoxy genistein and prunetin besides lupanone, lupeol and sitosterol. Leaves contain protein (14%), proanthocyanidins and phenolic acids such as *p*-hydroxybenzoic and gentisic acids. Flowers yield chalcones and aurones such as butin, butein and their glycosides, butrin, isobutrin, palasitrin, coreopsin, isocoreospin, sulphuretin monospermoside and isospermoside. Also present in flowers are waxes and β-sitosterol. Pods yield an imide, palsimide and 1-carbomethoxy, 2-carbamyl hydrazine. The seeds contain an imidazole alkaloid, monospermin, phytolectin and palasonin, apart from protein (20%), fatty oil (15.5%), mucilage (47%), β-sitosterol and α-amyrin (Guha et al. 1990; Sharma et al. 1991).

Butea gum is a powerful astringent, used for diarrhoea, dysentery, phthisis and haemorrhage from stomach and bladder. Bark is an alterative, aphrodisiac, anthelmintic and used for tumours, bleeding piles and ulcers. The decoction of bark is useful in menstrual disorders, cold and coughs and as a tonic. Roots are used in elephantiasis, night blindness and other problems of vision. Leaves are also tonic, diuretic and aphrodisiac and used to cure boils, pimples and tumorous haemorrhoids. Flowers also possess similar properties and are also used as an emmenagogue and for leprosy, leucorrhoea and gout. Seeds are used as vermifuge in veterinary medicine.

4.14 Antioxidant Therapy

Free radicals are a necessary evil. They are fundamental to any biochemical process and form an integral part of aerobic life and metabolism. They are generally produced by respiration, which uses oxygen and some cell immune functions. In the body, these compounds are counteracted and balanced by the natural antioxidants. According to the present knowledge, in addition to the traditional role of protecting the fats, proteins and carbohydrates, the biological antioxidants manage repair systems such as iron transport proteins (transferrin, ferrutin, caeruloplasmin, etc.), antioxidant enzymes and factors affecting signal transduction, vascular homeostasis and gene expression (Frankel and Mayer, 2000). But environmental pollutants such as air/water contaminants, radiation, pesticides, etc., produce a large influx of free radicals in the body and tip the balance between pro-oxidant (free radicals) and anti-free radical (antioxidicants) in favour of the former, resulting in a cumulative damage of protein, lipid, DNA, carbohydrates and membranes leading to oxidative stress. The oxidative stress, in which the free radicals outweigh antioxidants in number, is suggested to be the cause of ageing and other diseases like atherosclerosis, stroke, diabetes, cancer and neuro-degenerative diseases such as Alzheimer's disease and Parkinsonism. Studies conducted worldwide have provided enough evidence stating that providing the body systems with sizable amount of antioxidants may correct the vitiated homeostasis (Tiwari, 1999; Pietta, 2000) and a proper administration of antioxidants can treat or prevent the onset of ailments.

We normally receive a good amount of antioxidants through our diet consisting of spices, vegetables, pulses and cereals which contain a large variety of these compounds. Their role in preventing human diseases, including cancer, atherosclerosis, stroke, rheumatoid arthritis, neurodegeneration and diabetes are currently being followed (Fang et al. 2002). But at the time of oxidative stress, a good amount of external antioxidants are to be pumped into the body, and this is the basis of various antioxidant-based therapeutics.

Atherosclerosis is considered mainly due to oxidation of low-density lipoproteins (LDL) leading to the accumulation of cholesterol in the atherosclerotic lesion. In addition, the high fat diet also is found to decrease the expression of genes which produce free-radical scavenging enzymes resulting in an oxidative stress (Sreekumar et al. 2002). Recently, probucol and α-tocopherol-based compounds have been designed and developed to combat atherosclerosis, which are undergoing clinical trials. Similarly, therapeutic formulations from Arjuna, Abana, Shosaikoto, etc., having antiatherosclerotic properties are of great promise (Inoue, 2000; Wasserman et al. 2003). Stroke and cerebral thrombosis are the other related diseases receiving attention in free radical scavenging and antioxidant-based therapeutics.

The increased levels of sugar in blood, caused by **diabetes** also leads to the production of free radicals leading to oxidative stress. This paves ways to further cellular damage and late diabetic complications. In this case also, antioxidants are found to decrease oxidative stress and further complications (Karazu et al. 1997). Supplements of antioxidants have shown to decrease oxidative stress and diabetic induced defects in diabetic animals. Since some of the plants used for diabetes are rich sources of antioxidants, the activities of these plants are now considered based on their antioxidant principles (Scartezzi and Speroni, 2000).

Oxidative stress also is found to induce neuronal damage, leading to neuronal death and cause neurodegenerative diseases such as ***Alzheimer's and Parkinson's diseases***. In the former case, the characteristic histopathological alterations such as neutristic plaques composed largely of amyloid β-peptides and neuronal aggregates of abnormally phosphorylated cytoskeletal proteins are found to be due to the overproduction of ROS (Reactive Oxygen Species). The free-radical mediated selective degeneration of dopaminergic neurons in the nigrostriatal system, which results in loss of dopaminergic influence on other structures of basal ganglia, is found to result in normal Parkinsonian symptoms. A number of studies show that antioxidants like L-DOPA, tocopherols, ascorbate, etc., can block neuronal death and may also have therapeutic properties.

Cancer is another disease in which antioxidants can play a beneficial role. Metabolic activation of a carcinogen is a free radical-dependent reaction. It is also found that endogenous DNA damages arise from a variety of intermediates of oxygen reduction and free radicals and this plays a positive role in carcinogenesis. Misrepair of damaged DNA as also the oxidation of guanine by free radicals are other problems here. The higher amount of H_2O_2 and other reactive oxygen species in cancer cells also points to the persistent oxidative stress in cancer. Phenolics and isothiocyanates are effectively used in inducing H_2O_2 and they activate stress genes, which in turn, cause large-scale irreparable damage to DNA and thus inhibit proliferation and act as anticancer agents (Loo, 2003). The anticancer activities of antioxidants involve: (i) trapping the carcinogen; (ii) blocking the metabolic activation of carcinogens; (iii) scavenging the free radicals and their production; (iv) block lipoxigenase/cycloxygenase pathway, etc. (Smith et al. 1995).

From among the kinds of radicals (hydroxyl, alkoxyl, peroxyl and carbon centred) involved in oxidative stress, the peroxyl radicals are considered to be the major target radical for radical scavenging antioxidants in vivo (Niki *et al.* 1995). Peroxyl radicals are electrophilic in nature and, therefore, electron-donating substituents in phenolics increase the reactivity towards these radicals. Phenolics having substituents on the ortho-position are found to be better peroxyl scavengers. In addition, flavonols are found to contain a strong nucleophilic centre that reacts with electrophilic species and thereby decreases the bioavailability of the ultimate carcinogens.

In plants, phenolics form a very large group. They include simple phenolics, phenolic acids, phenyl propanes, acetophenones, stilbenes, xanthones, flavones, flavonols, catechins, aurones, chalcones, isoflavonoids, flavans, neoflavonoids, tannins, etc. All these types of phenolics are very effective antioxidants. The large number of medicinal plants containing phenolics (as active principles) will effectively support the role of these compounds in therapy. In addition, the various phenolics found in other medicinal plants (as such there is no plant without phenolics) having different active principles, will ably support the curative action of the latter compounds by providing a radical-free environment.

5

Gums and Mucilages

Gums and mucilages include all the hydrocolloids obtained from plants and are polysaccharides consisting of more than one type of monosaccharide residues. The gums are considered as pathologic products, produced in response to injury by a process known as "gummosis", whereby the cell walls and their ingredients are dissolved to form a colloid, which serves as a protective layer over the wounded tissue and later occurs as exudates from the various plant parts, especially the trunk. Mucilages, on the other hand, are classified as natural plant products produced by the plant for the imbibition and retention of water. But from a chemical point of view, gums and mucilages are almost identical and it is nearly impossible to draw a line demarcating one from the other.

The solutions of gums and mucilages are laevorotatory. On hydrolysis, they yield sugars like arabinose, galactose, glucose, mannose and xylose along with various uronic acids and methyl sugars. The sugar acids, when present in appreciable amounts, tend to lower the pH of the solution enabling the gums to occur frequently as salts of sodium, potassium, calcium or magnesium. In some cases, the sugar components are methylated (gum Tragacanth) or acetylated (Karaya gum). The trace amounts of nitrogen (0.08-5.6%), at times encountered in certain samples of gum, are considered to be due to be presence of proteins or sugar amines like glucosamine.

Gums containing linear polysaccharides are found to be less soluble in water, producing very viscid solutions. They tend to precipitate in course of time because of the inter-molecular hydrogen bonding facilitated by the parallel alignment of molecules and form colloidal gels-sols-possessing low surface tension and, therefore, act as important protective colloids and stabilizing agents. Even though the monomer components of most of the gums are known, the data on detailed structure are wanting in many cases. The great difficulty in isolating the polysaccharide components is the factor responsible for this situation. The final structure, when attained, may prove to have much relevance to the taxonomy of the group concerned.

Echinacea angustifolia DC. (Asteraceae)

Coneflower, Echinacea

Echinacea is a herbaceous perennial with a tapering, slightly spiral longitudinally furrowed root, conical heads with purple ray florets and yellowish purple disc florets on a high cone and 4-angled achenes. It is a native of Europe and America.

Roots and rhizome contain an arabinogalactan, inulin, a variety of phenols of which echinacoside is the principal component, cichoric acid, oil, resin, phytosterols and betaine. The polysaccharides are more abundant in leaves.

Echinacea is said to have immune stimulating properties. It is also an alterative and aphrodisiac. It is recommended for colds, coughs and flu for which the polysaccharides inulin and arabinogalactan are said to be responsible.

Plantago ovata Forsk. (Plantaginaceae)

Syn. *Plantago ispaghula* Roxb.

Isabgol, Psyllium

This is a woolly succulent nearly stemless annual with a basal rosette of linear acuminate toothed 3-nerved leaves, and flowers borne on ovoid or cylindric spikes. Capsules are ellipsoid with circumscissile dehiscence containing 2 seeds. The mucilaginous seeds are elliptical, boat-shaped with a reddish brown elongated spot on convex surface. A native of Sind, Isabgol is now cultivated in northwest India.

Seeds contain mucilage (20-30%), a fatty oil (2.5%), proteins and iridoids. The mucilage consists of weakly acidic arabinoxylans with a xylan backbone and branches of arabinose, xylose, rhamnose and galacturonic acid.

Psyllium is a bulk-forming laxative useful in constipation and diarrhoea. It is also found to be hypocholesteremic.

Hygrophila auriculata Heine (Acanthaceae)

Syn. *H.spinosa* T. Anders.
Asteracantha longifolia Nees.

Kokilaksha/Taalmakhana, Hygrophila

A stout armed, hispid herb with quadrangular stems, leaves in a whorl around the node of which two on the opposite side are long and oblong-lanceolate and the 4 smaller leaves (two on each side) bearing axillary sharp yellow spines, purplish-blue flowers in 4 pairs in each node and linear-oblong pointed capsules enclosing 4-8 hygroscopic, white hairy seeds on retinacula. A common weed in moist areas, *H. auriculata* is a native of South Africa.

Leaves contain lupeol, stigmasterol and flavonoids like apigenin and acacetin, proanthocyanins and phenolic acids, vanillic and syringic acids. Flowers yield apigenin glucuronide. Seeds contain mucilage and sterols.

Kokilaksha (seeds) is a reputed remedy for arthritis and is a well-known aphrodisiac. It is also used to cure oedema, ascites, thirst, bladder stones and is a blood purifier. It is also useful in cancer and tubercular fistula. Leaves and roots are taken for leucorrhoea. The whole plant is antihepatotoxic.

Sterculia spp. – (Sterculiaceae)

Karaya Gum, Sterculia Gum

The dried gummy exudates obtained from various species of *Sterculia*, indigenous to India, are known as Karaya gum. These trees attain a height of 10-15 m. The exudation of gum takes place by natural cracks or incisions.

The gum occurs in irregular colourless or pale yellow, translucent striated masses. It possesses an odour of acetic acid and is least soluble in water. Chemically, it is a partially acetylated polysaccharide of highly branched structure with residues of D-glucuronic acid (occurring as end groups), D-galacturonic acid, D-galactose and L-rhamnose and acetic acid. (Aspinall & Fraser, 1965). The galacturonic acid and rhamnose units are believed to be the branching points within the molecule.

On absorption of water, karaya gum swells up several times its original size and so is used as a bulk laxative. The other important uses are in textile printing and in preparation of building material.

Cochlospermum gossypium De Candolle (Malvaceae, **Kumbi)**, yields a gum often erroneously called as Karaya gum. This gum also contains the same sugar residues as karaya gum, viz., D-galactose, L-rhamnose, D-glucuronic acid (occasionally with its 4-methyl ether) and D-galacturonic acid, but differs from it in the nature of branching points and in the detailed sequence of sugars. It is used as a substitute of Karaya gum.

Anogeissus latifolia Wall. (Combretaceae)

Dhawa, Gum Ghatti

This tree, a native of India and Sri Lanka, is about 30 m high with small elliptic leaves, green apetalous flowers in dense globose heads and small compressed coriaceous samaroid fruits.

Gum ghatti, a brownish yellow solid, on hydrolysis yields mannose, galactose, arabinose and glucuronic acid. The known data suggest that the macromolecule is highly branched with long chains of 1-6, linked β-D-galactose with a small proportion of 1-3-links. The side chains contain glucuronic acid, 1→2 linked with mannose and/or 1→6 linked with galactose. The amount of arabinose is very high (40%) and it is proposed that these units occur as end groups and presumably form a surrounding sheath to the macromolecule (Percival, 1966).

Gum ghatti is used as a substitute for gum Acacia.

Abelmoschus esculentus Moench. (Malvaceae)

Syn. *Hibiscus esculentus* Linn.

Okra/Bhindi, Ladies finger

This stout rough annual is a native of tropical Africa, grown for the edible fruits used as a vegetable. The plant possesses cordate palmately lobed leaves, crimson-centred yellow flowers and horn-like angular hairy mucilaginous pods containing many round seeds.

The immature fruits contain carbohydrates (6.7%) and fats and are a good source of calcium, iron and other minerals. Glycosides of 3', 4'-diOMe quercetin, proanthocyanidins and hyperin also are reported (Daniel, 1989). Seeds contain fats (20%), proteins (20%), vitamin E. and traces of gossypetin and cyanidin. Okra mucilage is an acidic polysaccharide consisting of galactose (80%), rhamnose (10%), galacturonic acid (6%) and arabinose (3%). Certain samples contained protein, minerals and glucosamine.

The mucilage from the pods is said to have emollient and demulcent properties and is useful in dysentery. A decoction of fresh unripe capsule is given in cases of gonorrheal cystitis and urethritis. Seed extract is found to inhibit cancerous cell growth in vitro. The fruit is said to be an aphrodisiac. Tender pods are eaten in spermatorrhoea. The ripe seeds are roasted and used as a substitute of coffee. Seed flour is added to maize flour in Egypt and consumed. Okra mucilage is used as a plasma replacement or blood volume expander.

Abutilon indicum Sweet. *(Malvaceae)*

Atibala, Country Mallow

A. indicum is a shrubby tomentose perennial with simple long-petioled broadly ovate-cordate leaves, small yellow flowers in axillary solitary or terminal panicles and globose/hemispherical schizocarp with 14-20 stellate-hairy mericarps enclosing black reniform seeds. This plant is a common weed throughout India and is highly variable. A number of subspecies such as subsp. *indicum*, subsp. *albascens* and subsp. *guineense* have been recognized.

The whole plant, which is used as the drug, is reported to contain β-sitosterol, phenolic acids such as vanillic, caffeic, *p*-coumaric and *p*-hydroxybenzoic acids, amino acids and sugars. The leaves contain mucilage, glycosides of gossypetin, β-sitosterol and vitamin C. Flowers are found to have cyanidin and gossypetin glycosides. Seeds yield a mucilage (6%), protein (19%) and oil (5-14%- containing oleic acid and linoleic acids as major components).

Atibala root is considered as an aphrodisiac, diuretic and a nervine tonic. It is also used in piles, leucoderma and against worms. The whole plant (both in Ayurvedic and Unani medicines) is used as febrifuge, anthelmintic, anti-inflammatory and in urinary and uterine discharges. The leaves are cooked and eaten for piles and diarrhoea and a decoction is useful as a wash for gonorrhoea and other vaginal infections. Seeds are employed for their action against cough, gonorrhoea, gleet and chronic cystitis.

A. indicum is a source of a stem fibre also.

Abutilon hirtum Sweet.

Syn. *Sida hirta* Lam.

Bala

This is a highly viscid shrub with falcate stipules, yellow solitary flowers with a purplish centre, globose densely stellate schizocarp having 20-25 mericarps and stellate pubescent seeds.

This plant is used as a substitute of *A. indicum.*

Other species of *Abutilon* used in medicine are (1) *A. pannosum* Schlecht. (Syn.*A. glaucum* Sweet), the leaves of which are used as a cure for piles, (2) *A. ramosum* Guill. & Perr. exhibiting spasmolytic and CNS depressant properties and (3) *A. theophrastii* Medic. (Syn. *A. avicenniae* Gaertn), the seeds are used in dysentery, fistulae and eyesores.

Acacia senegal Willd. (Mimosaceae)

Syn. *A. verek* Guillem. &Perr.

Gum Arabic Tree

This is a small thorny tree/shrub of 4-6m ht, with yellowish-white bark, small bipinnate leaves with stipular thorns (leaflets 6-10), yellow flowers in 5-10 cm long spikes and flat pods containing 5-7 seeds. *A. senegal is* a native of North Africa, common in drier parts of Rajasthan, Gujarat and Haryana.

The gum is collected by making transverse incisions on the bark. The gum is acidic in nature and is a mixture of two polysaccharide fractions. The major sugar components are galactose (37%), arabinose (30%), rhamnose (11%) and glucuronic acid (13%). The roots contain betulin, sitosterol and ceryl alcohol. Bark yields uvaol, β-amyrin and sitosterol. Pods contain kaempferol, quercetin, β-amyrin, β-sitosterol and its glycoside.

Gum Arabic is used as a binding agent for pills. Roots of *A. senegal* are used against dysentery, gonorrhoea and nodular leprosy. The stem bark extract is found to be anti-inflammatory and spasmolytic.

Ulmus rubra Muhl. (Ulmaceae)

Slippery Elm

This native of Canada and the United States is a lofty tree. The branchlets are often scabrous and the bud scales brown tomentose. The ovate-oblong leaves are rough and unequal at the base. Flowers are borne in dense clusters and the fruit is a samara.

The inner parts of the bark yield mucilage, which is found to be a mixture of two or more polyuronides, mostly polygalacturonides.

The mucilage is a good demulcent and possesses great nutritive properties.

Alcea rosea Linn. (Malvaceae)

Syn. *Althea rosea* Cav.

Hollyhock

A. rosea is an erect, sparingly branched, stellately hairy annual/biennial reaching a height of 2m, with sub-orbicular palmately lobed leaves, variously coloured large flowers borne singly in axils of leaves and hirsute capsule splitting into orbicular mericarps. This is a native of China, cultivated in many gardens.

All parts of the plant contain mucilage. The flower colour is due to anthocyanins such as cyanidin glycosides or aromadendrin glucosides or flavonols such as herbacetin, kaempferol or quercetin glycosides. Seeds yield mucilage, fixed oil (9-18%), volatile oil (consisting of limonene, phellandrene and citral) and β-sitosterol. Alkaloids also are found in seed samples.

The roots of *A. rosea* are used in fever and dysentery. Flowers are used as cooling, emollient and diuretic as well as for rheumatism. Seeds are used as demulcent, febrifuge and diuretic.

Althea officinalis Linn. (Malvaceae)

Khaira, Marshmallow, Althea

A velvety-tomentose perennial reaching upto 2m ht, *A. officinalis*, possesses woody rootstock bearing 30 cm long corky roots, obovoid, ovate (lower) or lobed (upper) leaves, rose flowers in axillary clusters and a schizocarpic fruit. A native of Europe, this plant is grown in many places in India for its beautiful flowers and medicinal roots.

Mucilage is present in all parts of plant but is highest in roots (30%). It is an acidic polysaccharide consisting of galacturonic acid, galactose, glucose, xylose and rhamnose. Besides mucilage, the roots contain asparagine (2%), betaine, lecithin, phytosterol, starch (30%), pectin (11%) and small amounts of tannins. Leaves contain tiliroside, astragalin, isoquercitrin, hypoletin, vanillic, ferulic, caffeic, *p*-hydroxy benzoic and *p*-coumaric acids besides mucilage (15%) and a volatile oil. The flowers yield populnin, kaempferol, quercetin, luteolin, and 5% of mucilage and traces of a volatile oil. Seeds yield fatty oil (15%), rich in linoleic and oleic acids and a small amount of malvalic acid.

The roots, as well as the other parts of the plant, are used as demulcent, emollient and expectorant. The roots, as a poultice or fomentation, are used in inflammatory tumours, burns and other affections. Leaves, young tops and roots are used in salads. The flowers are a good expectorant.

Bambusa arundinacea Roxb. (Poaceae)

Syn.*B. bambos* Druce

Bansa, Bamboo

This is a spinous stately bamboo with short, stout, knotty rhizomes, dense hollow yellow culms reaching upto 30m in height, linear lanceolate leaves, large panicles and oblong 5-8 mm long grain (caryopsis) grooved on one side. The bamboo plant can be seen in moist parts of India, wild and cultivated.

Leaves contain crude protein (18.5%). Grains contain carbohydrates (76%), protein (13%) and more than 1% minerals. Roots contain benzoic acid and a cyanogenic glucoside, taxiphyllin. Internodes contain **tabashir.**

The stems and leaves are used in leucoderma, inflammations, bronchitis, gonorrhoea and fever. The leaves are considered emmanagogue, anthelmintic, astringent and febrifuge. Seeds are used as food.

Tabashir

Banslochan/Vanslochana

This is a form of amorphous silica occurring as pale blue/white translucent/transparent tasteless masses (2.5cm thick) in the hollow internodes of *Bambusa arundinacea, B.vulgaris, Dendrocalamus strictus* and *Melocanna baccifera.* This is mainly consisting of silicic acid (SiO_2, 97%) with 1% organic matter and traces of iron, calcium, alum and alkalies. It is mainly used as a tonic and aphrodisiac. Other uses include in asthma, cough, paralysis and general debility.

Basella alba Linn. (Basellaceae)

Syn.*B. rubra* Linn.

Upodika/Poi, Red Vine Spinach

B. alba is a succulent twiner with red or green stem, simple broad-ovate fleshy leaves, small pinkish white flowers borne on peduncled spikes and subglobose purple fleshy utricles. A native of tropical Asia, it is now widely cultivated as a pot herb.

The leaves contain protein (3%), carbohydrates (4%), betacyanins, carotenoids, oxalic acid, flavonoids such as acacetin, 7,4'-diOMe Kaempferol and 4'OMe isovitexin and phenolic acids like vanillic, syringic and ferulic acids. Fruits contain betacyanins, gomphrenin I,II & III (Magalan et al. 1991; Glassgen et al. 1993).

B. alba is used for digestive disorders, fistulae, pustules, inflammatory tumours and syphilitic ulcers in the nose. The leaf juice is a laxative, diuretic, and useful in gonorrhoea, balanitis and in catarrhal affections. The plant is said to exhibit antiviral properties. The dye from the fruits is used as a food colouring.

Benincasa hispida Cogn. (Cucurbitaceae)

Syn. *B. cerifera* Savi.

Koshmanda, Ash Gourd

B. hispida is gregarious climber having angular hispid stems, large 5-7 lobed leaves, axillary tendrils, unisexual large yellow flowers (monoecious) in the axils of leaves and a fleshy succulent elliptic cylindric fruit with a white removable waxy covering and white ovoid compressed seeds. A native of Indo-Malaya, this plant is cultivated throughout Asia.

Ash gourd contains only 3.5% dry edible matter. It is found to contain free amino acids including γ-aminobutyric acid and serine, sugars, mannitol, triacontanol, lupeol, β-sitosterol, adenine, trigonelline and histidine. The seeds yield a fatty oil (20%) rich in linoleic (63%), oleic (20%) and palmitic (10%) acids and protein (22%). Also present in the seeds are avenasterol, 24β-ethylcholesta-7, 25-dienol, euphol, tirucallol, cycloartenol, taraxasterol, α- and β-amyrins, butyrospermol, isomultiflorenol and multiflorenol. Leaves also are found to contain free amino acids and organic acids (Ramesh et al. 1989; Faure and Gaydou, 1991).

The fruit is styptic, laxative, diuretic and help to cure internal haemorrhages and diseases of the respiratory tract. Medicinal preparations of the fruit are recommended for epilepsy, constipation, piles, dyspepsia, syphilis and diabetes. Seeds and oil are anthelmintic and used for appendicitis.

Amorphophallus campanulatus Blue. (Araceae)

Suran, Elephant Foot Yam

A stout herb having an underground spherical corm, solitary tripartite leaf having a dark green warted petiole and obovate acute segments. Spathe is bell-shaped, greenish-pink enclosing a columnar cream-coloured spadix bearing unisexual flowers. A native of eastern India, now widely cultivated for the edible tubers.

The fresh corm is found to contain starch (18%), proteins (1.2%), oxalic acid (1.3%), protease inhibitors trypsin and chymotrypsin and minerals. Other compounds reported are an active diastatic enzyme, betulinic acid, β-sitosterol, stigmasterol and some free sugars. The irritant nature of corm is due to calcium oxalate.

The corm is used as a vegetable. The main medicinal property is in the treatment of dysentery, piles and haemorrhoids. It is also used as a stimulant, expectorant and to increase appetite. The roots are used in ophthalmia and as emmenagogue. The seeds and corm are used externally as an irritant to treat rheumatic swellings.

Ceratonia siliqua Linn. (Caesalpiniaceae)

Carob, Locust Bean

C. *siliqua* is a moderately tall, evergreen dioecious tree with paripinnate leaves bearing 3-5 pairs of elliptic coriaceous leaflets, white/yellow/green/red scented apetalous flowers (majority unisexual, but with a few perfect flowers), curved slightly compressed turgid reddish brown pods containing 3-6 shining ovoid seeds embedded in pulp. A native of Mediterranean region, carob is now widely cultivated.

The pods consist of pulp (90%), and seeds (8%). The pulp is one of the richest sources of sugars (upto 50%). The young pods are found to contain (+) catechin, (-) epicatechin, (-) epicatechingallate, phloroglucinol, gallic acid, ferulic acid, quercetin and myricetin. The ripe pods contain catechol tannins and prodephinidin. Seeds are rich in proteins (16%) and contain five C-glycosyl flavones such as three di-C-glycosyl apigenins (schaftoside, isoschaftoside and neoschaftoside) and two O-glucosides of schaftoside and isoschaftoside) along with campesterol, stigmasterol, β-sitosterol, avenasterol and α-amino pimelic acid. The gum, present in endosperm as a vitreous layer on the inner side of the seed coat, is a galactomannan. This is a branched polysaccharide consisting of a linear chain of mannose with galactose molecules attached to every 4th or 5th mannose residues. Oil from the seed germ contains linoleic (44%), oleic (38%) and palmitic (14%) acids (Batista & Gomes, 1993).

The pods are used as candy and as a sweetening agent in pharmaceutical preparations. They are tonics, expectorants and improve the voice as well as acting as antioxidants. The gum is employed in confectionery and in baby food for prevention of diarrhoea and in eliminating dehydration and maintaining electronic balance. It is also used as an excipient for tablets and for lubricating jellies, laxatives, lotions, etc.

Macrocystis spp. (Laminariaceae)

Algin

Algin or sodium alginate is the sodium salt of alginic acid, a polysaccharide obtained from brown seaweeds, mainly *Macrocystis pyrifera* Ag., native of the coasts of the USA and Canada. These "giant kelps" which grow in water 10-30 m in depth, possess repeatedly branched stipes reaching a length of 30-50 m. New blades are continuously formed at the apex of the branches at regular intervals and the mature blade possesses an elongate gas bladder at its base facilitating the floating of the upper portions of the plant in water.

Alginic acid, which forms the principal cell wall component, is extracted by macerating plant parts with dilute sodium carbonate solution, diluting with water, filtering and then precipitating with diluted Sulphuric acid or with $CaCl_2$. The free alginic acid is then converted to its sodium salt by neutralizing with Na_2CO_3.

Sodium alginate is a tasteless and odourless yellowish powder readily soluble in water, whereas alginic acid is relatively insoluble in water. Alginic acid is a linear copolymer of D-mannuronic acid and L-guluronic acid. These two types of monomers occur in three different block structures-homopolymeric blocks of D-mannuronic acid residues, homopolymeric blocks L-guluronic acid and blocks with an alternating structure (1-4 linked). The ratio of mannuronic acid to guluronic acid varies from 3:1 to 1:3, depending upon the source plants. This variation may be due to the relative amounts of varying block structures (Haug, 1974).

The alginates, especially sodium alginate, are extensively used in the food industry (mainly for ice creams) as a smoothening agent, sizing of paper, cosmetics and as a tablet binder and thickening agent. Calcium alginate (alginate fibres) is used as a haemostatic dressing or as a swab for pathological work.

Gelidium cartilagineum Gaillon (Rhodophyceae)

Agar

Agar is the hydrocolloid obtained from the various red algae like *Gelidium cartilagineum* Gaillon and other species of *Gelidium* (in Japan) or *Gracilaria confervoides* Grevil or *Petrocladia* sp. (in the United States and Australia). The agar is extracted by boiling the algae with acidulated water, filtering and cooling when the agar separates out as a jelly. The traces of water are removed by freeze-drying. Agar is manufactured only in winter.

Agar occurs in the form of translucent yellowish-white odourless flakes, stripes or powder. It is a heterogeneous polysaccharide containing two components, agarose and agaropectin. Agarose is a neutral galactose polymer consisting of 1-3 linked β-D-galactopyranosyl (some methylated at 6) and 1→4 linked 3, 6-anhydro-α-L-galactopyranosyl units. The structure of agaropectin is poorly understood, but is considered to be a polymer similar to agarose, in which the anhydrogalactose units are partly esterified with sulphuric acid. Agaropectin contains a few molecules of glucuronic acid and 1% pyruvic acid in acetal linkage to D galactose residues.

The most important use of agar is as gel in culture media. It is also used as an emulsifier, in cosmetics and paper industries and in medicine as a laxative.

Chondrus spp. and *Gigartina* spp. (Rhodophyceae)

Carrageenin, Irish Moss

Carrageenin is a group of hydrocolloids obtained from various red algae like *Chondrus crispus* Stackhouse, *Gigartina mamillosa* J. Agardh or related species, native to coasts of Ireland and the United States. The algae growing on rocks are collected, bleached with the help of sun and dew (sometimes with the help of SO_2), dried and stored.

The dichotomously branched thalli contain at least four polysaccharide complexes. They are galactans like agar, but contain D-galactose instead of L-sugar and higher content of sulphate esters. Carrageenins, on treatment with 0.25 KCI solutions, yield an insoluble fraction and a soluble fraction. The insoluble fraction is designated as K-carrageenin having an alternate structure of 3, 4-linked β-D galactose and 3, 6-anhydro α-D-galactopyranosyl (1→4)-3, 6, anhydro-D-galactose) and contains more than one sulphate ester group per disaccharide unit-probably at 6-position of D-galactose. The polymer chain, at times, is branched at every tenth D-galactose residue and the terminal non-reducing units are believed to be 3, 4- or 3, 6-disulphated galactose molecules. The KCI soluble fraction yields soluble and insoluble components on alkali treatment. The alkali soluble polysaccharide, λ-carrageenin is found to be a highly sulphated galactan usually with a low content of 3, 6-anhydro D-galactose. This polymer contains equal amounts of β-1, 3-and α-1, 4-linked galactose units with the 4-linked units 6-sulphated or 2, 6-disulphated. The alkali-insouble fraction is found to be identical with *k*-carrageenin, considered to be a precursor of the latter and is named *μ*-carrageenin. The precursor of λ-carrageenin is not yet isolated but designated *l*-carrageenin (Haug, 1974).

It is interesting to note that the sporophytes of *Chondrus crispus* contain mainly λ-carrageenin and the gametophytes *k*, and μ-carrageenins (McCandless et al. 1973).

Carrageenins are used extensively in the manufacture of cosmetics and as an emulsifying agent in medicine.

Part B
Primary Metabolites

6

The Primary Metabolics

Primary metabolites include carbohydrates, proteins and lipids. These compounds are believed to be continuously synthesized and utilized and are never of any adaptational significance as the secondary metabolites. But many primary metabolites are being stored in various tissues without any particular role and they can then be considered as secondary metabolites.

6.1 Carbohydrates

Carbohydrates are the major (in quantity) compounds in the plant world and form the largest share of combined carbon in plants. They are polyhydroxy aldehydes or ketones or their derivatives and are the main source of energy for the plant.

Based on the sugar units present, carbohydrates are classified into monosaccharides, disaccharides and polysaccharides. Monosaccharides are further divided to monoses, bioses, trioses, etc., depending on the number of carbons they possess. The number of monosaccharides known is about sixty. Glucose and xylose are the most common sugars in plants followed by galactose, arabinose and mannose. Sugars like hamamelose are present in certain tannins only. The sugars present in cardiac glycosides are unique in the sense that they are not present anywhere else in the plant kingdom.

Sucrose and maltose form the most common disaccharides in nature. Sucrose is present naturally in sugarcane, sugarbeet, sugar maple and in many palms. Invert sugar (an equimolecular mixture of glucose and fructose formed by the action of invertase on sucrose) is abundant in honey. Lactose is the milk sugar present in fruits like *Ceratonia*, *Achras* and saponins of *Sapindus* and gentiobiose is a glycosidic component of amygdalin and some cardiac glycosides. Sophorose is the sugar component of stevioside and rutinose is of rutin. Sambubiose and vicianose are components of many glycosides.

Oligosaccharides like tri-/tetrasaccharides are present in saponins and cardiac glycosides. Gentianose occurs free in gentian roots, whereas planteose is present in *Psyllium*. Lychnose and maltotetrose are the tetrasaccharides common in members of Caryophyllaceae and Liliaceae, respectively.

Sugar alcohols, sugar acids, sugar amines also abound in the plant kingdom. Ribitol amounts to 4% in *Adonis* whereas mannitol amounts to 6%

in *Randia* bark. Sorbitol is widespread in fruits of Rosaceae and dulcitol is common in Celastraceae. *myo*-Inositol is considered as a vitamin by many. Bornesitol, polygalitol, quebrachitol, etc., are some other sugar alcohols reported from many plants.

Among the sugar acids, L-ascorbic acid (vitamin C) is the most distinguished, being the antiscorbutic factor and an effective antioxidant present in Emblica and Citrus fruits. Glucuronic, galacturonic and mannuronic acids are components of many gums and mucilages and glycosides.

The organic acids form another group of compounds related to carbohydrate metabolism, widely distributed in higher plants. These acids may be mono-, bi- or tricarboxylic acids. The tricarboxylic acids occur in catalytic amounts in all tissues, but only citric and malonic acids regularly accumulate in plant tissues. Citric acid is common in many of the fruits such as orange, lemon, strawberry and gooseberry, whereas the dominant acid in grape, plum, apple and cherry is malonic acid. Acetic acid, being a universal precursor of fatty acids and related lipids, is present in all plants. Monofluoroacetic acid, a toxic analogue of acetic acid, is reported from a South African plant *Dichapetalum cymosum*.

Among the polysaccharides, starch, pectin and cellulose are the most common in plants. Dextrins, prepared by the partial hydrolysis of starch, are used as the filling material in tablets and capsules. Inulin, the polymer of fructose, present in many Compositae tubers like *Helichrysum, Inula*, etc., serves as the food for diabetic patients. The gums and mucilages as well as pectin, with their high viscosity, serve as laxatives. Many polysaccharides of natural origin are used as viscosing agents in drug delivery systems of ophthalmic preparations. Of late, polysaccharides are receiving a lot of attention for their discovered biological activities. It has been found that the polysaccharides (TS-polysaccharide) isolated from *Tamarindrus indicus* exhibited a protective role in reducing UV-B derived damage to eye cells (Raimondi et al. 2003). Polysaccharides of Hyssop were found to inhibit SF strain of HIV-I viurus.

Gums and mucilages are considered secondary metabolites and are treated separately.

"Dietary fibre" includes all unassimilable structural plant materials like pectin, cellulose, hemicelluloses, lignin, etc. This group has important effects on gut function due to their bulk, the ability to absorb water, absorptive and combining properties and their being substrates for normal bacteria of the gut. Various ailments like constipation, diverticular diseases and appendicitis are correlated with low intake of dietary fibre.

6.2 Aminoacids and Proteins

The plant world possesses more than 300 amino acids, 20-26 being the components of proteins and rest non-protein amino acids. Among the protein amino acids, deficiency of essential amino acids causes malnutrition. Non-protein amino acids, more than half, are restricted to the order Fabales (Leguminoseae) occur free or as peptides and exhibit a variety of biological/pharmacological properties.

There are seven classes of non-protein amino acids:

(1) *D-Amino acids*: They are enantiomers of protein amino acids and are discovered from many plants. They act as inhibitors of plant metabolism and a similar role can be expected in animals too. D-Leucine, D-phenylalanine and D-glutamic acids are found to be the components of certain antibiotics while D-proline is present in ergot alkaloids.

(2) *Non-α-amino acids*: They contain amino groups in any carbon other than α-carbon. γ-Aminobutyric acid and β-alanine are widely distributed in plants and are also reported from a few medicinal plants.

(3) *Protein amino acid derivatives*: They are the derivatives of protein amino acids. DOPA (dihydroxy phenyl alanine) reported from *Mucuna* seeds exhibit distinct medicinal properties. Canavanine and N-N-dimethyl tryptophan are some other members of this group. Ornithine is an intermediate in tropane and pyrrolidine biosynthesis.

(4) *New heterocyclic and alicyclic amino acids*: The homologues of proline, azetidine-2-carboxylic acid (4 membered heterocyclic ring) and pipecolic acid (6-membered heterocyclic ring) are present in many plants. Hypoglycin A is a cyclopropane amino acid reported from *Blighia* and *Litchi*, exhibiting marked hypoglycemic activity. Mimosine and lathyrine are toxic components of certain pulses.

(5) *Dicarboxylic acids and amides*: α-Amino adipic acid and α-amino pimelic acid are two higher homologues of glutamic acid widespread in higher plants.

(6) *Sulphur containing amino acids*: A number of sulphur containing amino acids are present in many members of Liliaceae, Brassicaceae and Leguminosae. They are especially abundant in species of *Allium*. Djenkolic acid is a derivative of cysteine obtained from djenkol beans and *Pithecellobium*.

(7) *Hydroxy amino acids*: γ-Hydroxy glutamic acid is a widely distributed member of this group. Serotonin (5-hydroxytryptamine) is the irritative component of *Mucuna* fruit hairs.

Low molecular weight peptides are a group grossly neglected by all. They may go unnoticed in a survey in the sense that they look like common amino acids in chromatograms. A number of dipeptides are seen in the plant kingdom. Glutathione is a tripeptide, first obtained from yeast, found to be widely distributed in plants and acts as a prosthetic group in some enzymes. Recently it has been found to detoxify some components of kava extract (Whilton et al. 2003). Glutathione, as an immune system enhancer, is present in fairly good quantities in asparagus, broccoli, cabbage, cauliflower, potato, tomato, avocado, grapefruit, oranges, peaches and watermelon. Toxins of mistletoe and the alkaloids of some members of Rhamnaceae, Rubiaceae and Urticaceae are true peptides. Canthiumine is one such tetrapeptide isolated from *Canthium* (Rubiaceae) and scutiamine is a hexapeptide obtained from *Scutia* (Rhamnaceae). Here it is worth mentioning that peptides of fig were found to inhibit Angiotensin-1-converting enzyme (ACE).

Heteromeric or conjugated peptides (containing a non-amino acid component) can also be seen in nature. Pantothenic acid of coenzyme A consists of pantoic acid and β-alanine. Folic acid (a member of vitamin B complex) contains pteroic acid linked with L-glutamic acid. Lathyrus factor is γ-L-glutamyl-β-amino propionitrile which is a neurotoxic factor present in *Lathyrus* seeds. The peptide alkaloids of ergot as well as ziziphine and related alkaloids (from *Ziziphus*) contain lysergic acid and pyrocoll, respectively, in addition to peptides.

Among proteins, some lectins (glycoproteins, phytohaemagglutinins) are important. Though they were once considered toxic, they are being reinvestigated for their interaction with protoplasm. Abrin, the toxalbumin from Abrus as well as ricin from Castor are studied extensively for their anticancer properties.

Citrus spp. (Rutaceae)

Citrus is a genus of evergreen aromatic shrubs/small trees with axillary thorns, winged petiole, gland-dotted leaves, fragrant white flowers and fruit in the form of a hesperidium. The various species useful are *C. limon* Burm.f (lemon), *C. medica* Linn. (citron), *C. paradisi* Macfad. (grape fruit), *C. reticulata* Blanco (orange) and *C. sinensis* Osbeck. (mosambi). All of them provide refreshing drinks from *fruits* and are a source of numerous medicines.

All *Citrus* fruits are rich in vitamin C (ascorbic acid) and minerals. The constituents of fruits are citric and malic acids, their salts, sugars, essential oils, pectins, anthocyanins, limonoids and carotenoids. *Citrus aurantifolia* contains maximum amount of citric acid (7%) whereas this plant and *C. sinensis* contain maximum ascorbic acid (30-58mg/100g).

All these fruits are rich in free amino acids, which amounts to about 70% of the total nitrogenous compounds. All parts of these plants contain coumarins such as bergaptene, bergemotin, bergaptol, scopoletin, imperatorin and psoralens. The fruits, especially the rind, contain flavonoids such as flavones, flavanones and anthocyanins. The flavanones such as naringin and neohesperidin are bittering agents useful in tonics and beverages. *Citrus* flavones can be converted to dihydrochalcones, which are sweetening agents. Limonoids are the compounds providing bitterness to many *Citrus* fruits. Limonin, the most common limonoid, is the principal bitter principle. Limonoids tend to disappear when the fruit ripens, being converted to non-bitter derivatives. Seeds contain glucosides of limonin, obacunone, deacetyl nomilin, nomilin, ichangin, isolimonic acid, etc.

All *Citrus* fruits, being sources of vitamin C, are used for treating scurvy. The rinds of fruits, especially of orange, are the sources of bioflavonoids, which, once designated vitamin P (permeability factors), are excellent antioxidants and strengthen capillary walls.

Emblica officinalis Gaertn. (Euphorbiaceae)

Syn. *Phyllanthus emblica* Linn.

Amalaki/Amla, Indian Gooseberry

E. officinalis is a tall deciduous tree with small narrowly oblong leaves arranged distichously on branches so that the young branches resemble a pinnately compound leaf, small green unisexual flowers and depressed globose, obscurely six-lobed fleshy fruits containing six (sometimes only one) trigonous seeds. It is a native to India, now growing wild as also cultivated.

The fruit contain vitamin C (ascorbic acid, upto 2%, richest source) gallotannins (5%), carbohydrates (14%), phenolic acids, alkaloids, pectin and minerals. The phenolic acids include gallic, ellagic and phyllemblic acids and emblicol. Also present are a good amount of free aminoacids such as alanine, aspartic acid, glutamic acid, lysine and proline The alkaloids present are phyllantidine and phyllantine. Curcuminoids also are reported from certain samples.

Emblica exhibits antitumour, hypotensive and immunomodulatory properties. Extracts are anticytotoxic, antigenotoxic, anticlastrogenic, and antimutagenic in nature. This is one of the richest and stablest sources of vitamin C because of the presence of antioxidant gallotannins in the fruit. The drug is a *rasayana* drug. It is used in the treatment of peptic ulcer and dyspepsia.

Portulaca oleracea Linn. (Portulacaceae)

Lonika, Purslane

This is a prostrate/erect succulent herb with green/purplish branchlets, subopposite/alternate spathulate fleshy leaves, bright yellow flowers (sepals two, stamens many) crowded at the tips of branches and an ovoid circumscissile capsule enclosing many black concentrically striate granulate curved seeds. A pantropical weed.

The whole plant, especially leaves, is exceptionally rich in antioxidant vitamins A, C and E. Also present are betacyanins, glutathione, minerals, oil containing gamma linolenic acid, oxalic acid and phenolic acids, vanillic and syringic acids.

Purslane is considered a tonic, cooling, blood purifier and useful in cough, inflammations of bladder and kidney and piles. The whole plant is used as a vegetable.

Tamarindus indica Linn. (Caesalpiniaceae)

Cinca, Tamarind

T. indica is a large tree with paripinnate leaves having 15-17 pairs of small subsessile oblong leaflets, creamy yellow (pink dotted) zygomorphic flowers in axillary racemes (stamens only 3, monadelphous) and an oblong incurved thick torulose pod containing black shining obovate seeds enveloped by a fleshy sweet acidic mesocarp. A native of tropical Africa, tamarind is now cultivated widely in the tropics.

The brownish red pulp surrounding the seed contains soluble sugar (30%), pectin and tartaric acid (12-15%). More than half of the tartaric acid present is in the form of potassium hydrogen tartarate. Leaves contain tartaric and malic acids, flavones like apigenin, acacetin and luteolin, glycoflavones such as the vitexin, isovitexin, orientin and isoorientin and phenolic acids vanillic, p-coumaric, *p*-hydroxy benzoic and gallic acids. Also present are prodelphinidins (Daniel, unpublished).

The fruit pulp is carminative, digestive, laxative, antiscorbutic and antibilious. Tender leaves and flowers are antibilious and cooling and used for constipation, dyspepsia, flatulence and urinary infection. Infusion of leaves is used as a gargle for sore throat and for cleaning ulcers. Flowers are used as a paste for eye diseases.

Oxalis corniculata Linn. (Oxalidaceae)

Cangeri/Indian Sorrel

A small procumbent acidic herb with a creeping stem, rooting at nodes, palmately 3-foliolate leaves (leaflets obcordate, emarginated at the apex), small yellow flowers in axillary umbellate clusters and a linear oblong capsule containing white arillate seeds. A cosmopolitan plant.

Leaves contain tartaric acid and citric acids, calcium oxalate, flavones (acacetin and 7,4'-diOMe apigenin), glycoflavones (4'-OMe vitexin, 4'-OMe iso- vitexin and 3',4'-diOMe orientin), flavonols (3',4'-diOMe quercetin) and phenolic acids such as *p*-hydroxybenzoic, vanillic and syringic acids. (Umadevi et al. 1988).

The whole plant is a household remedy for indigestion and diarrhoea. It is also useful in dysentery, piles, fever and biliousness. Paste of leaves used externally to remove warts, corns and also applied to inflamed parts.

Garcinia indica Choiss. (Clusiaceae)

Syn. *G.purpurea* Roxb.
Brindonia indica Dup.

Kokam, Brindon

This is a small tree with glabrous oblong-lanceolate leaves (red when young) and solitary or fascicled greenish-white unisexual flowers. Male flowers are tetramerous with many stamens, female flowers are with staminodes. The fruit is a purple berry containing 5-8 compressed seeds embedded in pulp. Kokam is grown in coastal areas of the Western Ghats.

The Kokam seed is the source of kokam butter. The fruit rind contains camboginol and benzophenone derivatives, garcinol and isogarcinol. Also present in the fruit are (+) hydroxycitric acid and cyanidin glycosides.

The fruit of this tree is antiscorbutic, cholagogue, emollient and demulcent. Its oil is used in skin diseases.

Garcinia mangostana Linn.

Mangusta, Mangosteen

This is a small laticiferous tree yielding the edible fruit mangosteen. It bears elliptic-lanceolate glossy leaves, solitary pink flowers and globose, slightly flattened reddish berries with a persistent thick calyx at the base, a thick rind and 5-8 flat seeds enclosed in a thick white jelly like pulp.

The pulp contains carbohydrates (14%) mainly invert sugar and sucrose, citric acid (0.5%), pectin and tannins (7-14%). Bark, fruit hulls and dried latex yield xanthones, mangostin, β-mangostin, and γ-mangostin (normangostin). Fruits contain gartenin, 8-deoxygartenin, normangostin and garcinones A-C. Maclurin is also reported from the fruits (Balasubramanian and Rajgopalan,1988).

Mangostin and other xanthones are found to be hepatoprotective, CNS depressants, anti-inflammatory and antiulcer. Mangostin is a cardiotonic.

Garcinia combogia Desv.

A moderate-sized tree resembling *G. indica*, but differing in a grooved (7-8) berry, each vertical furrow containing an ovoid compressed seed surrounded by a white aril. *G. combogia* is common in the South Western Ghats.

The latex yields cambogin and camboginol, two isoprenylated benzophenones; the fruit rind contains (-)- hydroxycitric acid (about 30%).

The fruit rind is used as a spice and believed to help in slimming. The plant is used in skeletal fractures. The dried rind is used in rickets and enlargement of the spleen.

Garcinia morella Desv.

Ursina-Gurgi, Gamboge

G.morella differs from other spp. in having a smaller fruit (less than 2.5 cm diam). The fruit is 4-lobed, enclosing 4 ovoid-reniform compressed seeds. Gamboge is common in south India.

The latex collected from the bark forms "gamboge". This contains morellic and isomorellic acids. Seed coat yields morellin, isomorellin, isoneomorellin and α,β,γ and x- guttiferins. The heartwood contains morelloflavone.

Gamboge is used as a hydragogue, cathartic and possesses anthelmintic and cathartic properties. It is also useful in cerebral affections, amenorrhoea and as an anti-inflammatory agent.

6.3 Fatty Acids, Triacylglycerols and Glycolipids

Many of the saturated and unsaturated fatty acids present in plants serve as storage food, energy reserves and components of phospholipids. Linoleic acid, when present, is converted in the body (through *gamma*-linolenic acid) to prostaglandins, which are hormone-like chemical messengers regulating body functions such as smooth muscle contraction, blood pressure control and responses to inflammation. Imbalances or deficiencies of essential fatty acids in the body cause many disorders such as asthma, migraines, inflammations, diabetes, arthritis and other metabolic disorders (Foster, 1997). A number of uncommon fatty acids also are present in certain plants, which exhibit significant medicinal properties. Seeds rich in gamma linolenic acid as in evening primrose are useful in premenstrual syndrome, diabetes, obesity and heart diseases. Ricinoleic acid (12-hydroxy oleic acid) present in castor oil is responsible for the purgative action of the oil. Cyclopentene acids such as chaulmoogric and hydnocarpic acids are present in *Hydnocarpus*. Glycolipids found in turpethum are another example of fatty acid derivatives useful in medicine.

Alkanes, fatty alcohols and other derivatives are abundant in cutins and suberins. Hydroxy fatty acids, present in plants like *Clitoria*, form lactones (aparajitin), while those present in conifers form polymeric estolides. The pharmacological properties of these compounds are not known. Being lipid soluble, they are always present in an oil extract of the plant.

Oenothera biennis Linn. (Onagraceae)

Evening Primrose

This is a tall biennial herb, with stem having red blotches, crinkled sagittate leaves, yellow flowers (4 petals) opening in evenings, and elongated capsules. A native of North America, evening primrose is now widely cultivated in gardens.

Seeds yield oil rich in essential fatty acids such as *cis*-linoleic acid, alpha-linoleic acid and gamma linoleic acid. Also present in the oil is Vitamin E.

Evening primrose oil is recommended for PMS (premenstrual syndrome) like irritability, breast pain, tenderness and moodchanges. The oil is also useful in eczema, blood pressure and rheumatoid arthritis. The flowers, leaves and stem bark are used for their astringent, sedative and expectorant properties. The roots are used for haemorrhoids.

Hydnocarpus wightiana Bl. (Flacourtiaceae)

Garudaphala, Chaulmoogra

Chaulmoogra oil is obtained from the seeds of this native of India. The plant is a tall dioecious tree with pubscent branches, elliptic leaves, unisexual white solitary or racemed flowers and an apple-sized tomentose globose berry enclosing numerous obtusely angular seeds within a hard rind.

The oil consists of a mixture of glycerides of hydnocarpic (45%), chaulmoogric (20%), gorlic (15%), oleic (12%) and palmitic acids (Hydnocarpic and chaulmoogric acids possess cyclopentene ring systems). The pericarp contains propelargonodin. Seed hulls yield flavanolignans, hydnocarpin, isohydnocarpin and methoxyhydnocarpin. Seed coat contains other flavanolignans hydnowightin and neohydnowightin (Parthasarathy et al. 1979).

The unsaturated fatty acids show strong bactericidal properties towards the Micrococcus of leprosy and so are used for the treatment of leprosy.

Ricinus communis Linn. (Euphorbiaceae)

Erandah, Castor

Castor plant is a tall, soft, hollow-stemmed glabrous laticiferous shrub with long-petioled peltate palmately 7-9 lobed, (lobes serrate) leaves, monoecious inflorescences containing male apetalous flowers with branched stamens at the top and female flowers (also apetalous) towards the base. Fruit is a 3-valved spinous schizocarp (regma) containing 3 mottled and carunculate seeds. A native of tropical Africa, castor is now extensively cultivated for oil.

Castor seed contains oil (45%) and protein (16%), enzymes like lipase, alkaloid ricinine and allergens. Ricin, a toxalbumin, is the principal haemagglutinin. Ricinine ($C_8H_8O_2N_2$, 1-methyl-3-cyano-4-methoxy-2-pyridone), a water-soluble alkaloid, is present mostly in the seed coat. The oil contains mainly ricinoleic acid (upto 90%). Leaves contain ricinine, N-methyl ricinine and glycosides of kaempferol and quercetin

Roots and leaves are considered highly effective in rheumatism. They are used for dyspnoea, hydrocele, flatulence, dysentery, piles, leprosy and arthritis. Seed oil is purgative and used for fever, inflammation of scortum and colic. Tender leaves are used in jaundice.

Exogonium purga Benth. (Convolvulaceae)

Jalap

Jalap is the resin obtained from *E. purga*, a Mexican twiner having thin horizontal underground runners from the nodes of which the tuberous roots arise.

The dried roots yield about 10% resins, a volatile oil, starch, gum and sugars. The resin contains convolvulin as the principal component, which on hydrolysis, yields glucose, rhodeose and convolvulinic acid. Convolvulinic acid is assumed to be an ester of 11-hydroxy tetradecanoic acid (convolvulinolic acid) and five glucose molecules. Also present in the resin are ipurganol (a phytosterol glycoside), and fatty acids such as palmitic, stearic, valeric, tiglic and exogonic acids.

Jalap is a drastic purgative and hydragogue.

Ipomoea orizabensis Led. (Convolvulaceae)

Orizaba Jalap

A native of Mexico, this twining perennial produces a fusiform root of 60 cm long.

The roots contain 6-18% glycosidal resins, volatile oil, scopoletin and fatty acids. Jalapin, the principal component of resin, yields on hydrolysis jalapinolic acid (11-hydroxy palmitic acid), glucose, rhodeose and methyl tetrose. Sitosterol and other phytosterol glycosides are the minor components of this resin.

Used as cathartics and hydragogues.

Operculina turpethum Silva Manso (Convolvulaceae)

Syn. *Ipomoea turpethum* R.Br.

Trivert, Indian Jalap

This is a large woody laticiferous twiner with 4-angled winged twisted stem, long slender fleshy much-branched roots, ovate-oblong long-petioled leaves having a cordate base, creamy white flowers in axils of large lanceolate bracts, borne in corymbose cymes and a globose circumscissile capsule, enclosed in a woody calyx, containing 4 black seeds. A native of India, this plant is a common twiner in tropics of India, Africa and Australia.

The resin, which amounts to 10% in the roots, contains three glycosides turpethin (major component), α-turpethin and β-turpethin and a volatile oil.

Root and root bark are used as a purgative in case of constipation, piles and jaundice. It is also successfully employed in cases of dropsy due to heart, kidney or liver diseases. This plant is recommended for rheumatic and paralytic affections and for various types of inflammations.

General References

Anonymous (1955-1992) *Wealth of India* (Raw Materials) including revised volumes of A, B, C-Ci. Publication and Information Directorate, CSIR, New Delhi.

Anonymous (1992) *Second Supplement to Glossary of Indian Medicinal Plants with Active Principles, Part I* (A-K). Publication and Information Directorate, CSIR, New Delhi.

Anonymous (1998) *Indian Herbal Pharmacopoeia* Vol. **I** RRL, Jammu & IDMA, Mumbai.

Anonymous (1999) *Indian Herbal Pharmacopoeia* Vol. **II** RRL, Jammu & IDMA, Mumbai.

Anonymous (2000) *The Wealth of India, First Supplementary Series (Raw Materials)* Vol. **1 A-Ci**. NISC, CSIR, India.

Anonymous (2001) *The Wealth of India, First Supplementary Series (Raw Materials)* Vol. **2 Ci-Cy.** NISC, CSIR, India.

Anonymous (2002) *The Wealth of India, First Supplementary Series (Raw Materials)* Vol. **3 D-I.** NISC, CSIR, India.

Chopra, R. N. et al. (1982) *Chopra's Indigenous Drugs of India.* Academic Publishers, Kolkata.

Duke, J. (1997) *The Green Pharmacy,* (Indian Edn.) Scientific Publishers, Jodhpur, India.

Latha S. and M. Daniel (2001) *Journal of Food Science Technology* **38(3):** 272.

Robinson, T. (1975). *The Organic Constituents of Higher Plants,* Cordus Press, Mass.

Sabnis, S. D and M. Daniel (1990) *A Phytochemical Approach to Economic Botany,* Kalyani Publishers, New Delhi.

Sivarajan, V. V. and I. Balachandran (1994). *Ayurvedic Drugs and their Plant Sources,* Oxford & IBH, New Delhi.

Tiwari, A.K. (2004) *Current Science,* **86(8):** 1092.

Wagner, H and P. Wolff (1977) *New Natural Products and Plant Drugs with Pharmacological, Biological or Therapeutical Activity* (Ed.) Springer–Verlag, Berlin.

References

Abe, F., Yamauchi, T. and A. S. C. Van, (1988) *Phytochemistry* **27:** 3627.

Abe, F., Van, A. S. C. and T. Yamauchi, (1989) *Chemical and Pharmaceutical Bulletin* **37:** 2639.

Abdel Kadar, M. S., Omar, A.A., Abdal-salam, N. A. and F. R. Stermitz. (1994) *Phytochemistry* **36:** 1431.

Abeyasekara, I. (1990) *Fitoterapia* **61:** 473.

Ahmed, B. and C. Yu (1992) *Phytochemistry* **31:** 4382.

Ahmed, V., Perveen, S. and S. Bano. (1989) *Planta Medica* **58:** 474.

Ahmed V., Parveen S. and S. Bano (1990) *Phytochemistry* **29**: 3287.

Ahmed V., Fizza, K., Aziz-Ur Rahman and S. Arif (1987) *Journal of Natural Products* **50**: 1186.

Akiyama T., Tanaka, O. and S. Shibata (1972) *Chemical and Pharmaceutical Bulletin* (Tokyo) **20:** 1957.

Anjanayelu A. S. R., Sastry, C. S. P. Narayan, G. K. and L. R. Rao (1978) *Journal of Indian Chemical Society* **55:** 1169.

Ansari A. A., Kenne, L., Atta-Ur-Rahman and T. Wehler (1988a) *Phytochemistry* **27:** 3979.

Ansari A. K., Aswal, B. S., Chander, R., B. N. Dhawan., Garg, N.K., Kapoor, N. K., Kulshreshta, D. K., Mehdi, H., Mehrotra, B. N., Patnaik, A. K., and S. K. Sharma (1988b) *Indian Journal of Medical Research* **87:** 401.

Appendino K. Tagliapiatra, S., Nano, G. M. and J. Jakupovic (1994). *Phytochemistry* **35:** 183.

Aruna K. and V.M. Sivaramakrishnan (1996) *Phytotherapy Research*. **10:** 577.

Aspinall, G.O. and R.N. Fraser (1965) *Journal of Chemical Society*, 4318.

Asthma R. K., Sharma, N. K., Kulshreshtha, D. K. and S. K. Chatterjee (1991) *Phytochemistry* **30:** 1037.

Asselieiah L. M., Hernandez O. and J. R. Sanchez (1990) *Phytochemistry* **29**: 3095.

Bagchi A., Hikino, H. and Y. Oshima (1991) *Planta Medica* **57:** 282.

Bagchi A., Oshima Y and H. Hikino (1991b) *Planta Medica* **57:** 96.

Balasubramanian K and K Rajagopalan (1988) *Phytochemistry* **27** : 1552.

Bakuni D. S and J. Sudha (1995) Medicinal and Aromatic plants. In *Advances in Horticulture*, Chadha K. G. and R. Gupta (Eds) Malhotra Publishing House, New Delhi Vol. **II**: 115.

Barua A. K., Pal S. K. and S. P. Dutta (1972) *Journal of Indian Chemical Society* **49:** 519.

Barua A. K. Chakraborti, P., Gupta, A. S. D., Pal, S. K., Basak, S., Banerjee, S. K. and K. Basu (1976) *Phytochemistry* **15:** 1780.

Batista M. T. and E. T. Gomes (1993) *Phytochemistry* **34:** 1191.

Begerhotta A. and N. R. Banerjee (1985) *Current Science* **54:** 690.

Bessale R. and D. Lavic (1992) *Phytochemistry* **31:** 3648.

Bhutani K. K., Vaid, R. M., Ali, M., Kapoor, R., Soodan, S. R. and D. Kumar (1990) *Phytochemistry* **29**: 969.

Bisset, N. G. (1994) *Herbal Drugs and Phytopharmaceuticals*. Medpharm Scientific Publishers, Stuttgart pp. 303.

Blasko A., Sheih, H., Pezzuto, J. M. and G. A. Cordell (1989) *Journal of Natural Products* **52:** 1363.

Bradley P. R. (1992) *British Herbal Compendium*: British Herbal Medicine Association, Dorcet, Vol. **I**. 145.

Burks D. and M. Woods (1963). *Radiation Research Supplements* **3**: 212.

Carpenter I., Locksley, H. D. and F. Sheimann (1969) *Phytochemistry* **8:** 2013.

Chakravarty A. K., Mukhyopadhyaya, S., Masuda, K. and A. Ageta (1992) *Tetrahedron Letters* **33:** 125.

Chandra R., Deepak D. and A. Khare (1994) *Phytochemistry* **35:** 1545.

Choi Y., Hussain, R. A., Pezzato, J. M., A. D. Kinghorn and J. F. Mortan (1989) *Journal of Natural Products* **52:** 1118.

Connolly J. D. (1983) Chemistry of the Limonoids of the Meliaceae and Cneoraceae. In *Chemistry and Chemical Taxonomy of the Rutales*. Waterman, P. G. and M. F. Grundon (Eds) Academic Press, London, p. 175.

Crowden R. K. and S. J. Jarman (1974). *Phytochemistry* **13:**1947.

Cuendet, M. and J. M. Pezzuto (2004) *Journal of Natural Products* **67(2):** 269.

Dan K and D. Dan (1988) *Fitoterapia* **59:** 348.

Daniel M. (1989) *Current Science* **46(14):** 472.

Daniel, M and S.D. Sabnis (1977) *Current Science* **46(14):** 472.

Daniel, M and S.D. Sabnis (1978) *Indian Journal of Experimental Biology* **16(4):** 512.

Daniel, M and S. D. Sabnis (1978a) *Indian Journal of Experimental Biology* **16(4):** 512.

Daniel M. and S. D. Sabnis (1982) *Indian Botanical Reporter* **1(2)**: 84.

Daniel M. and S. D. Sabnis (1986) *Biome* **1(2):** 65.

Daniel, M. and S. D. Sabnis (1987) *Indian Botanical Reporter* **6(1):** 53.

Darwish F. M. M and M. G. Reionecke (2003) *Phytochemistry* **62(8):** 1179.

Das, B and K. Chakravarty (1991) *Indian Journal of Chem*istry **30B:** 1052.

Das P. C., Joshi P. C. and S. Mandal (1992) *Indian Journal of Chem*istry **31B:** 342.

Dayal R. (1988) *Journal of Scientific and Industrial Research* **47:** 215.

Deepak M. (1999) *Standardization of Some Herbal Drugs Using Chromatographic techniques*. Ph. D Thesis, BIT. Pilani, India.

Devon, T. K. amd A. I. Skott (1972) *Handbook of Naturally Occurring compounds: Terpenes*. AP: N.Y.

Dhar, K. L. and A. K. Kalla (1974) *Phytochemistry* **13(2):** 2894.

Dimitrova, Z., Lateva, S. and N. Dimitrova, (1993). *Acta Microbiologica Butig.* **29:** 65.

Dixit V. P., Singh, B., Dixit, V. D., Lohan, I. S. and G. C. Georgie (1992) *Phytotherapy Research* **6:** 270.

Dragull K., Yoshida W. Y. and C. Chang (2004) *Phytochemistry* **63(2):** 193.

Dreyer D. L. (1983) Limonoids of the Rutaceae. In *Chemistry and Chemical Taxonomy of the Rutales,* Waterman P. G. and M. F. Grundon (Eds) AP. London p. 215.

Elgamal M. H. A., Shalaby, N. M. M., Duddick, H. and M. Hiegemaan (1993) *Phytochemistry* **34:** 819.

Endo K., Kanno E. and Y. Yoshima (1990) *Phytochemistry* **29:** 797.

Erdtmann H. (1955) Tropolones In *Modern Methods of Plant Analysis,* Paech, K. and M. V. Tracey (Eds) Springer-Verlag, Berlin **3**: 351.

Fairburn, J. W. (1964). *Lloydia* **17**: 79.

Fang Y. Z., Yang S. and G. Wu, (2002) *Nutrition,* **18:** 872.

Fatope M. O., Suad Khamis S. Al-Burtomani, John O. O, Abdulrahman O. Abdulnour, Salma M. Z. Al-Kindy and Y. Takeda, (2003) *Phytochemistry* **62(8):** 1251.

Faure R and E. M. Gaydou (1991) *Journal of Natural Products* **54:** 1564.

Fossen T. and O. M. Anderson (2003) *Phytochemistry* **62(8):** 127.

Fozdar I., Akina, J. C. and B. Richardson (1989) *Phytochemistry* **28** : 2459.

Frankel E. N. and A. S. Mayer (2000) *Journal of the Science of Food and Agriculture* **80:** 1925.

Ghosal S., Kaur, R. and S. K. Bhattacharya (1988) *Planta Medica* **54:** 561.

Gibbs R.D. (1974) *Chemotaxonomy of Flowering Plants.* McGill Queens, Montreal.

Gijibels M. J. M., Seheffer J. J. C. and S. A. Baerheim (1982) *Fitoterapia* **53:** 17.

Glassgen W. E., Metzger, J. W., Hever, S. and D. Strack (1993) *Phytochemistry* **33:** 1525.

Glotter J. (1991) *Natural Products Reports* **8:** 415.

Goh S. H. and I. Jantan (1991) *Phytochemistry* **30:** 366.

Gollapudi, S. Sharma, H. A., Aggarval, S. Byers, L. D. and S. Gupta (1995) *Biochemistry and Biophysics Research Communications* **210(1):** 145.

Grue–Sorensen G. and K. Spenser, (1993) *Journal of American Chemical Society* **115:** 2052.

Guha P. K., Poi R. and A. Bhattacharya (1990) *Phytochemistry* **29:** 2017.

Gupta M. M. and R. K. Varma (1991) *Phytochemistry* **30:** 875.

Gupta M. M., Verma R. K. and L. N. Mishra (1992) *Phytochemistry* **31:** 4036.

Gupta V., Aggarval, A., Singh, J. and H. P. Tiwari (1989) *Indian Journal of Chemistry* **28B:** 282.

Guvanov I. A., Libizov, N. J. and A. S. Gladkikh, (1970) *Farmatsiya* **19:** 23.

Hagnauer R. (1966) Alkaloids In *Comparative Phytochemistry,* T. Swain (Ed.) A.P., London p.211.

Hamerski L., Furlan, M., Siqueira Silva, D. H., Cavalhiero, A. J., Eberlin, M. N., Tomazela D. M. and V. S. Bolzani (2003) *Phytochemistry* **63(4):** 397.

Handa, S. S., Singh, R., Maurya, R., Satti, N. K., Suri, K. A. and O. P. Suri (2000) *Tetrahedron Letters* **41:** 1579.

Handa S. S., Chawla, S. S., Sharma, A. K. and K. L. Dhar (1989) *Fitoterapia* **60:** 195.

Handa S. S. Sharma, A. K. and K. L. Dhar (1992) *Fitoterapia* **63:** 3.

Harborne J. B. (1964). *Phytochemical methods,* Chapman and Hall, London.

Harvey D. J. (1981) *Journal of Chromatography* **212:** 75.

Hatam N. A. R., Whiting, D. A. and N. J. Yousif (1989) *Phytochemistry* **28:** 1268.

Haug A. (1974) Align and related hydrocolloids. In *Biochemistry Series* I. Northcote D. H. (Ed). MTP International Review of Science, Butterworth, London.

Hayashi T. and T. Yamagishi (1988) *Phytochemistry* **27:** 3696.

Herzberg S. and S. Liaan Jensen (1971) *Phytochemistry* **10:** 3121.

Hikino H. (1985) Recent Research in Oriental Medicinal Plants. In *Economic and Medicinal Plant Research.* Wagner H., Hikino H and N. R. Farnsworth, (Eds), AP, London Vol. **I.**: 53.

Hisham, A., Virginia, D., Marks, R and J. W. Carpenter (1992) *Planta Medica.* **58:** 474.

Hong S., Hanquing, W., Yongquiang and C. Yaozu (1991) *Phytochemistry* **30:** 1547.

Hopkinson, S. M. (1969) *Quaternary Reviews* **23**:98.

Horif T., Tominaya H and Y. Kawamura (1993) *Phytochemistry* **32**: 1076.

Hosny M., Khalifa, T., Calis, J., Wright, A. D. and O. Sticher (1992a) *Phytochemistry* **31:** 340.

Hosny M., Khalifa, T., Calis, J., Wright, A. D. and O. Sticher (1992b) *Phytochemistry* **31:** 3565.

Imamura, K., Strutz, T., Takamasu, T. and N. Miura (1993) *Journal of Natural Products* **56(12):** 2091.

Inoue, M. (2000) *Drug News Prospect* **13:** 407.

Iinuma M. Tosa, H., Tanaka, T. and S. Yonemori, (1994) *Phytochemistry* **35:** 527.

Itoh A., T. Tanahashi and N. Nagakowa (1991) *Phytochemistry* **30:** 3117.

Jackson, D. E. and P. E. Dewick (1984) *Phytochemistry* **23:** 1147.

Jain D. C. (1987) *Phytochemistry* **26:** 2223.

Jain N., Alam, M. S., Kamil, M., Ilyas, M., Niwa, M. and A. Sakae (1990) *Phytochemistry* **29:** 3988.

Jash S. S., Biswas, G. K., Bhattacharya, S. K. Chakraborty, A. B. K. Chowdhury (1992) *Phytochemistry* **31:** 2503.

Joly, A., Haag-Berrurier, H. and R. Anton (1980) *Phytochemistry* **19(9):** 1999.

Jondiko K. and A. Pattenden (1989) *Phytochemistry* **28:** 319.

Joshi K and P. Dev (1988) *Indian Journal of Chemistry* **27B:** 12.

Kano Y., Tanabe M. and M. Yasuda (1991) *Medicinal and Aromatic Plant Abstracts* 1103.

Kanwar U. Batla, A., Sanyal, S. N. and A. Ranga et al. *Journal of Ethnopharmacology* **28:** 249.

Karazu, C. (1997) *Metabolism* **46:** 872.

Kawai K., Akiyama, T., Ogihara, Y. and S. Shibata (1974) *Phytochemistry* **13:** 2829.

Keville K. and M. Green (1995) *Aromatherapy: A Complete Guide to the Healing Art*, Crossing Press, USA.

Khan A. and A. Malik (1989) *Phytochemistry* **28:** 2859.

Kikuzaki H. and N. Nakatani (1996) *Phytochemistry* **43:** 273.

Kitajima, J., Ischikowa, T., and M. Satoh (2003) *Phytochemistry* **64(5):** 1003.

Kubota T., Kitani H. and H. Hinoh (1969) *Chemical Communications* 1313.

Kundu S., Subhash, C. S., Bagehi, A. and A. B. Ray (1989) *Phytochemistry* **28:** 1769.

Lamparsky D. and I. Klinies (1988) *Perfumer and Flavorist* **13:** 17.

Lakshmi, K and V. Krishnamoorthy (1991) *Indian Journal of Pharmaceutical Sciences* **53:** 94.

Laphookhieo, S., Cheenpracha, S., Karalai, C., Chantrapromma, S., Rat-a-pa, Y., Ponglimanont, C. and K. Chantrapromma, (2004) *Phytochemistry* **65(4):** 507.

Liu H., Cheng, T. C. E., Ng, C. T. and M. Yuan (1993) *Planta Medica* **59:** 376.

Loo G. (2003) *Journal of Nutritional Biochemistry* **14:** 64.

MacCandless E. L., Craigie J. S. and J. A. Watter (1973) *Planta* **112:** 201.

Mahesh K and A. Bhaumik (1987) *Indian Journal of Chemistry* **26B:** 86.

Mahmood J., Singh, P. and S. Bhargava (1988). *Planta medica* **54:** 468.

Mandal, S., Das, P. C., Joshi, P. C., Das, A. and A. Chatterjee (1991) *Indian Journal of Chemistry* **30B:** 712.

Mandal S. and A. Chatterjee (1987) *Tetrahedron Letters* **28**: 1309.

Mangalan, S., Daniel, M and S. D. Sabnis (1988) *Indian Journal of Botany* **11(2):** 206.

Mangalan S., Daniel M. and S. D. Sabnis (1991) *Biologica Indica* **2 (1):** 23.

Markham K. R., Wallace, J. W., Babu, Y. N., Murthy, V. K. and M. G. Rao (1989) *Phytochemistry* **28:** 299.

Materska M. Piacente, S., Stochmal, A., Pizza, C., Oleszek, W. and I. Perucka (2003) *Phytochemistry* **63(8):** 893.

Maurya R., Wazir, V., Kapil, A. and R. S. Kapil (1996) *Natural Products Letters* **8:** 7.

Maurya R. Singh, R., Deepak, M. M., Handa, S. S., Yadav, P. P. and P. K. Mishra (2004) *Phytochemistry* **65(7):** 915.

Maurya R., Dhar K. L. and S. S. Handa (1997) *Phytochemistry* **44:** 749.

Moon C. K., Park N. S. and S. K. Koh (1977) *Chemical Abstracts* **87:** 114582.

Musago L. and G. Rodghiero (1972) Psoralens and their properties. In *Phytophysiology* Giesse, A. C. (Ed.) AP. New York **7:** 115.

Nagakura N., Itoh A. and T. Tanahashi (1993) *Phytochemistry* **32:** 761.

Nair, G. G., Daniel, M and S. D. Sabnis (1988a) *Current Science* **57(10):** 527.

Nair G. G., Daniel, M and S. D. Sabnis (1988b) *Geobios* **15:** 241.

Nawwar M. A. M., Et–Mousallamy M. D. and H. H. Barakat (1989) *Phytochemistry* **28** : 1755.

Niki E., Noguchi, N., Tsuchihashi, H and N. Gitoh (1995) *American Journal of Clinical Nutrition* **62:** 1322.

Oshiro M., Kuroyanagi M. and A. Ueno (1990) *Phytochemistry* **29:** 2201.

Okuda T., Yoshida, T. and W. Feng (1993) *Phytochemistry* **32:** 1033.

Pachey P. and C. Sehneidir (1981) *Archieves of Pharmacy* **314:** 251.

Parthasarathy, M. R., Ranganathan, K. R. and D. K. Sharma (1979) *Phytochemistry* **18:** 506.

Pathak V. P. and R. N. Khanna (1987) *Phytochemistry* **26:** 2103.

Pardhy R. S. and S. C. Bhattacharya (1978) *Indian Journal of Chemistry* **16B:** 171.

Patra A., Mishra, S. K. and S. K. Chaudhari (1988) *Journal of Indian Chemical Society* **65:** 205.

Pauson P. L. (1955) *Chemical Reviews* **55:** 9.

Pendse V. K., Dadhich, A. P., Mathur, P. M., Bal, M. S. and B. R. Madan (1997) *Indian Journal of Pharmacology* **9:** 221.

Percival E. (1966) The natural distribution of Plant polysaccharides. In *Comparative Phytochemistry* Swain T. (Ed.) AP. London p. 139.

Pietta P. G. (2000) *Journal of Natural Products* **63:** 1035.

Polonsky J. (1983) Chemistry and Biological Activity of the Quassinoids. In *Chemistry and Chemical Taxonomy of the Rutales* Waterman P. G. and M. F. Grundon (Eds) AP. London, p. 247.

Prabhakar P., Gandhidasan, R., Raman P. V., Krishnasamy N. R. and S. Nanduri, (1994) *Phytochemistry* **36:** 817.

Raimondi, L., Lodovici, M., Guglielmi F., Banchelli, G., Ciuffi, M., Boldrini, E and R. Pirisino., (2003) *Journal of Pharmacy and Pharmacology* **55:** 333.

Ramesh K., Akkina, J. C., Richardson, C. and C. Aquilera (1989) *Fitoterapia* **60:** 241.

Rastogi S., Pal R. and D. K. Kulshrestha (1994) *Phytochemistry* **36:** 133.

Reddy, M. K., Reddy, M. V. B., Gunasekar, D., Murthy, M. M., Caux, C. and B. Bodo (2003) *Phytochemistry* **62(8):** 1271.

Ren Y. J., Chen, H. S., Yang, G. J. and H. Zhu (1995) *Medicinal and Aromatic Plant Abstracts*: 1166.

Rükker G., Walter, R. D., Manns, D. and R. Mayer (1991) *Planta Medica* **57**: 295.

Rukker G., Walter, R. D., Manns, D and R. Mayer (1992) *Phytochemistry* **26:** 340.

Rucker G. Paknikar, S. K., Mayer, R., Breitmaier, E., Will, G. and L. Wiehl (1993) *Phytochemistry* **33:** 141.

Sajid T. M. (1996) *Phytotherapy Research* **10:** 178.

Satyanaryana V., Krupadanum, C. L. D. and G. Srimannarayanan (1991) *Indian Journal of Chemistry* **30B:** 989.

Scartezzini P. and E. Speroni (2000) *Journal of Ethnopharmacology* **71:** 23.

Shamsuddin T., Rahman, W., Khan, S. A., Shamsuddin, K. M. and J. P. Kindzinger (1988) *Phytochemistry* **27:** 1908.

Sharma S., Batra A. and B. K. Mehta (1991) *Indian Journal of Chemistry* **30B:** 715.

Sharma P. V. (1982) *Indian Journal of Pharmaceutical Sciences* **44:** 36.

Sharma S., Batra A. and B. K. Mehta (1987) *Indian Journal of Chemistry* **30B:** 715.

Sharma S. C., Chand R., Sati O. P. and A. K. Sharma (1983) *Phytochemistry.* **22:** 1241.

Sheeja M. J. (1991) *Chemical and Taxonomic Studies on the Malvaceae and Related Taxa.* Ph. D. Thesis, M. S. University, Vadodara.

Siddiqui B. S., Usmani, S. B., Begum, S. S. Siddiqui (1994) *Journal of Natural Products* **57:** 27.

Sikroia B. C., Srivasthava S. J. and G. S. Niromjan (1982): *Journal of Indian Chemical Society* **59:** 905.

Singh B., Saxena, S. K., Chandan, B. K., Agarwal, S. G., Bhatia, M. S. and K. K. Anand, (1984) *Indian Journal of Pharmacology* **16:** 139.

Singh P. and S. Bhargava (1992) *Phytochemistry* **31:** 2883.

Sinha K. (1986) *Advances in Biosciences* **5(1):** 69.

Skopp K. and H. Horster (1976) *Planta Medica* **29:** 20.

Smith T. J. (1995) *Annals of New York Academy of Sciences* **86(8):** 1092.

Sreekumar R., (2002) *American Journal of Physiology and Endocrinology Metabolism* **282:** E1055.

Sticher O., (1977) Plant Mono-, Di- and Sesquiterpenoids with Pharmacological or Therapeutical Activity. In *New Natural Products and Plant Drugs with Pharmacological, Biological or Therapeutical Activity,* Wagner P. and P. Wolff (Eds) Springer Verlag, Berlin, p. 137.

Stuppner H. and H. Wagner (1989) *Planta Medica* **55:** 467, 559.

Stuppner H. and H. Wagner, (1993) *Planta Medica (Supplement)* **59(7):** A583.

Sudhir S., Agarwal, S. K., Verma, S. and S. Kumar (1986) *Planta Medica* **I.:** 61.

Suresh A., Anandan J. and G. Sivanandan (1985) *Journal of Researches in Ayurveda and Sidha* **6(2):** 171.

Suzuki N., Miyase T. and A. Ueno (1993) *Phytochemistry* **34:** 729.

Swami K. D., Malik G. S. and N. P. S. Bisht. (1989) *Journal of Indian Chemical Society* **66:** 288.

Tanaguchi M., Yanai, M., Xiano, Y. Q., Kido, T. and K. Baba, (1996) *Phytochemistry* **42:** 843.

Tandon M. and Y. N. Shukla, (1993) *Phytochemistry* **32:** 1624.

Taneja R. and K. Tiwari (1975) *Tetrahedron Letters* 1995.

Thomas P. J. (1989) *Chemosystematic and Pharmacognostic Studies on Some Indian Members of the Rubiaceae* Ph. D. Thesis, M. S. University, Vadodara.

Thomson R.H. (1971) *Naturally Occurring Quinones* A.P. London.

Tiwari A. K. (1999) *Journal of Medicinal and Aromatic Plant Sciences* **21:** 730.

Tomoda N. Gonda, R., Shimizu, N., Kanari, M. and M. Kimura (1990) *Phytochemistry* **29:** 1083.

Triebs, W. (1953) London.P.211 *Sitzber Deutsch Academica Wissen Berlin Kl. Math u allgem. Naturw.*

Umadevi, I., Daniel, M. and S. D. Sabnis (1988) *Advances in Biosciences* **7(1):** 79.

Umadevi, I., Daniel, M and S. D. Sabnis (1990a) *Indian Journal of Botany* **13:** 23.

Umadevi I. and M. Daniel (1992b) *Acta Botanica Indica* **19:** 16.

Varljen K. (1989) *Phytochemistry* **28:** 2379.

Varshney I. P. and N. K. Dube (1970) *Journal of Indian Chemical Society* **47:** 717.

Verma R. K. and M. M. Gupta (1988) *Indian Journal of Chemistry* **27B:** 283.

Vogel G. (1977) Natural substances with effects on the liver. In *New Natural Products and Plant Drugs with Pharmacological, Biological or Therapeutical Activity,* Wagner P. and P. Wolff, (Eds) Springer Verlag, Berlin, p. 249.

Wadhwa V., M. Gupta, D. N., Singh, C. and V. P. Kamboj (1986) *Planta Med.* **3:** 231.

Wadhwa V., Singh, M. M., Gupta, D. N., Singh, C. and V. P. Kamboj, (1980) *Phytochemistry* **19:** 2040.

Wagner H., Wittman D. and W. Schater (1975) *Tetrahedron Letters* 547.

Wagner H., Greyer, B., Kiso, Y., Hikino, H. and G. S. Rao, (1986) *Planta Medica* **5:** 370.

Wagner H., Rao, G. S., Greyer, B., and H. Hikino (1984) *Planta Medica* **51:** 419.

Wasserman M. A., (2003) *American Journal of Cardiology (Supplement)* **91:** 34.

Wazir V., Maurya R and R. S. Kapil (1995) *Phytochemistry* **38:** 447.

Weidenfeld M. and K. Roder (1991) *Planta Medica* **57:** 578.

Weiss S. G., Tin-Wa, M., Perdue, R. E. and N. R. Farnsworth (1975) *Journal of Pharmaceutical Sciences* **64:** 95.

Whilton P., Lau, A., Salisbury, A., Whitehous, J. and C. S. Evans (2003) *Phytochemistry* **64:** 673.

Wohlbling R. H. and K. Leonhardt (1994) *Phytomedicine* **1(1):** 25.

Yadav P. P., Ahmad, G. and R. Maurya (2004) *Phytochemistry* **65(4):** 439.

Yadav S. K., Jain, A. K., Tripathi, N. and J. P., Gupta (1989) *Indian Journal of Medical Research* **90B:** 496.

Yamauchi T., Abe, F., Chen, R. F., Nonaka, G Santisuk, T and W. G. Padolina (1990) *Phytochemistry* **29:** 3547.

Yamauchi T., Abe, F., Padolina, W. G. and F. M. Dayrit (1990) *Phytochemistry* **29:** 3321.

Yan W., Ohtani, K., Kasai, R. and K. Yamasaki (1996) *Phytochemistry* **42:** 1417.

Yin M., Chen C. and Q. Wang (1988) *Chemical Abstracts* **108:** 164703.

Yoshikawa K., Arihara, S., Matsura, K. and Y. Miyase (1992) *Phytochemistry* **31:** 237.

Yoshikawa K., Matsura, K. and Y. Miyase (1994) *Chemical and Pharmaceutical Bulletin* **42:** 1226.

Zheng, G.Q., Kenny, P.M. and L. K. T Lam (1992) *Journal of Natural Products* **55:** 999.

Zepesochnaya G., Kurkin, V., Okhanov, V and A. Miroshnikov (1992) *Planta Medica* **58:** 192.

Appendix

Presented below are a few lists of the most useful herbs employed in various ailments. The plants recommended included a wide variety, distributed across almost all parts of the world. Special emphasis is given for Indian/ Ayurvedic plants. They can be easily administered in the form of a herbal tea or with milk.

1. General Tonics

1. *Panax ginseng*—Ginseng–Roots
2. *Withania somnifera*—Ashwagandha–Roots
3. *Phyllanthus emblica*—Emblica–Fruits
4. *Boerhavia diffusa*—Punarnava–Roots
5. *Tinospora cordifolia*—Amruta–Roots
6. *Chlorophytum borivilianum* and other species of Safed musali–Roots
7. *Asparagus adscendens*—Safed musali roots
8. *Curculigo orchioides*—Musali–Roots
9. *Rubia cordifolia*—Mangishta–Roots
10. *Hemedesmus indicus*—Sariva–Roots
    Ichnocarpus frutescens—Sariva–Roots
    Marsdenia tenacissima—Sariva–Roots
11. *Sida spinosa* and other species of *Sida, Abutilon, Urena,* etc., which are used as Bala
12. *Barleria prionitis* and related Sahachara–roots

2. Arthritis/Rheumatism

1. *Hygrophila auriculata*—Talmakhana–Seeds
2. *Zingiber officinale*—Ginger–Rhizome
3. *Gossypium herbaceum* and related species—Cotton Seeds
4. *Tinospora cordifolia*—Amruta–Roots
5. *Ricinus communis*—Castor–Roots
6. *Sida* spp./*Abutilon spp*.–Bala–Roots
7. *Withania somnifera*—Ashwagandha–Roots

8. *Vitex negundo*—Nirgundi–Especially useful for rheumatic swellings
9. *Argemone mexicana*—Seed oil as an external application
10. *Derris indica*—Seed oil as an external application

3. Diabetes

1. *Pterocarpus marsupium*—Vijayasar–Wood
2. *Eugenia jambolina*—Jamun–Seeds
3. *Gymnema sylestre*—Madhunashi–Leaves
4. *Enicostemma littorale*—Mamejav–Whole plant
5. *Trigonella foenum—graecum*–Fenugreek–Seeds
6. *Momordica charantia*—Bitter gourd–Fruits
7. *Coccinia grandis*—Bimbi–Fruits
8. *Catharanthus roseus*—Periwinkle–Leaves

4. High Blood Pressure

1. *Terminalia arjuna*—Arjuna–Bark
2. *Rauwolfia serpentina*—Sarpagandha–Roots
3. *Crtaegus* spp.—Hawthorn
4. *Allium sativum*—Garlic–Cloves
5. *Ammi majus*—Khella–Seeds
6. *Rubia cordifolia*—Mangishta–Roots/Stem
7. *Oenothera biennis*—Evening Primrose
8. *Zingiber officinale*—Ginger–Rhizome
9. *Vaccinium myrtillus* or any other source of anthocyanin including red grapes
10. Tea or any source of Bioflavonoids
11. *Apium graveolens*—Celery–Seeds
12. Any source of dietary fibre like carrot, fenugreek, beans, *Plantago*, etc.

5. Hypertension/Insomnia/Anxiety Neurosis or for Increasing Memory Power

1. *Bacopa monnieri*—Brahmi–whole plant
2. *Centella asiatica*—Brahmi–whole plant
3. *Acorus calamus*—Vacha–Rhizome
4. *Evolvulus alsinoides*—Shankapushpi–whole plant
5. *Convolvulus prostratus*—Shankapushpi–whole plant
6. *Canscora decussata*—Shankapushpi–whole plant
7. *Withania somnifera*—Ashwagandha–Roots
8. *Tinospora cordifolia*—Amrut–roots
9. *Ocimum tenuiflorum*—Tulsi–leaves

10. *Celastrus paniculatus*—Jyotishmati–seeds
11. *Ginkgo biloba*—Ginkgo–leaves
12. *Tinospora cordifolia*—Amrut–roots
13. *Hypericum perforatum*—St. John's wort

6. Female Urinogenital Problems

1. *Saraca indica*—Asoka, Bark
2. *Ficus benghalensis*—Vad–Bark
3. *Ficus religiosa*—Peepal–bark
4. *Woodfordia fruticosa*—Dhataki–flowers
5. *Symplocos cochinchinensis*—Lodhra–bark
6. *Asparagus racemosus*—Shataveri–root tubers
7. *Aloe barbadensis*—Aloe–Leaf mucilage
8. *Ficus racemosa*—Udumbar bark
9. *Mimosa pudica*—Lajjalu–whole plant
10. *Glycine max.*—Soybean–seeds
11. *Hibiscus rosa—sinensis*–Shoeflower–flowers
12. *Cissus qudrangula*—Cissus–whole plant
13. *Sida cordifolia*—Bala–root

7. Skin Care, Pimples, Ageing and Cosmetics

1. *Curcuma longa*—Turmeric–rhizome
2. *Curcuma zedoaria*—Karchurah–rhizome
3. *Rubia cordifolia*—Madder–roots
4. *Thespesia populnea*—Paras–bark
5. *Azadirachta indica*—Neem–bark, leaves
6. *Cassia fisula*—Aragvadhah–bark
7. *Cassia occidentalis*—Kasamardah–All parts, especially seeds
8. *Cassia tora*—Cakramardah–seeds
9. *Costus speciosus*—roots and rhizomes
10. *Ficus benghalensis*—Vad–bark, aerial roots
11. *Ficus religiosa*—Peepal–bark
12. *Derris indica*—Karanj–bark, seed oil
13. *Plumbago zeylanica*—Chitrak–roots
14. *Pterocarpus santalinus*—Red sandalwood–wood
15. *Pterocarpus marsupium*—Malabar kino–wood
16. *Psoralea corylifolia*—Bavchi–Seeds
17. *Calendula officinalis*—Calendula–whole plant
18. *Matricaria recutita*—Chamomile–whole plant
19. *Aloe barbadensis*—Aloe-leaves
20. *Centella asiatica*—Brahmi–whole plant

8. Hair Care

1. *Bacopa monnieri*—Brahmi–whole plant
2. *Centella asiatica*—Brahmi–whole plant
3. *Cardiospermum halicacabum*—Indravalli, whole plant
4. *Gmelina arborea*—Kasmari, fruits
5. *Eclipta alba*—Bangra, whole plant
6. *Phyllanthus emblica*—Emblica, fruits
7. *Terminalia bellerica*—Bahira, fruits
8. *Vitex negundo*—Nirgundi–Leaves
9. *Indigofera tinctoria*—Nili–whole plant
10. *Hibiscus rosa- sinensis*—Shoeflower–flowers

9. Obesity

1. *Holoptelia integrifolia*—Holoptelia–bark
2. *Garcinia indica*—Kokam–fruits
3. *Commiphora wightii*—Guggul–resin
4. *Boswellia serrata*—Frankincense–resin
5. *Averhoa bilimbi*—Bilimbi–fruits
6. *Moringa oleifera*—Sigru–fruits

10. Cough, Bronchitis

1. *Glycyrrhiza glabra*—liquorice–roots
2. *Adhatoda zeylanica*—Adhatoda–leaves
3. *Trachyspermum ammi*—Ajwain–fruits
4. *Coleus ambonicus*—leaves
5. *Piper nigrum*—Black pepper—fruits
6. *Piper longum*—Long pepper—fruits
7. *Ocimum tenuiflorum*—Tulsi–leaves
8. *Zingiber officinale*—Ginger–rhizome
9. *Curcuma longa*—Turmeric–rhizome
10. *Tylophora indica*—Tylophora–leaves (especially good for asthma)
11. *Allium sativum*—Garlic–cloves
12. All the aromatic spices of umbelliferae

11. Fever, Cold

1. *Caesalpinia bonduc*—Fevernut, seeds
2. *Andrographis paniculata*—whole plant
3. *Salix* ssp.—Willow–bark
4. *Zingiber officinale*—Ginger–rhizome
5. *Echinacea*—Feverfew

6. *Evolvulus alsinoides*—Shankapushpi–whole plant
7. *Vernonia cinerea*—Sahadevi–whole plant
8. *Oldenlandia corymbosa*—Oldenlandia–whole plant
9. *Capsicum* spp.—Red pepper–fruits
10. *Sambucus nigra*—Elderberry–fruits
11. *Ephedra sinica*—Ephedra–stem
12. *Allium sativum*—Garlic–cloves

12. Aphrodisiacs

1. *Mucuna pruriens*—Atmagupta–seeds
2. *Chlorophytum borivilianum*—Safed musali–tubers
3. *Asparagus adscendens*—Safed musali–tubers
4. *Hygrophila auriculata*—Talmakhana–seeds
5. *Panax ginseng*—Ginseng–roots
6. *Withania somnifera*—Ashwagandha–roots
7. *Ginkgo biloba*—Ginkgo–leaves/seeds
8. *Zingiber officinale*—Ginger–rhizome
9. *Aspidosperma* sp—Quebraco–bark
10. *Sida cordifolia*—Bala–bark
11. *Serenoa repens*—Saw palmetto–seeds
12. *Avena sativa*—Oats–grains

13. Diuretic, Urinary, Kidney Stones

1. *Boerhavia diffusa*—Punarnava–root
2. *Tribulus terrestris*—Gokhru–fruits
3. *Scoparia dulcis*—whole plant
4. *Aerva lanata*—whole plant
5. *Zingiber officinale*—Ginger–rhizome
6. *Equisetum arvense*—Horsetail–whole plant
7. *Mentha Piperita*—Peppermint–whole plant
8. *Silybium marianum*—Milk thistle
9. *Solidago virgaurea*—Golden rod–flowering tops
10. *Petroselinium crispum*—Parsley–root

14. Liver Problems

1. *Phyllanthus amarus*—Boiamla—whole plant
2. *Silybium marianum*—Milk thistle–seeds
3. *Boerhavia diffusa*—Punarnava–roots
4. *Terminalia chebula*—Haritaki–fruits

15. Galactogogues

1. *Polygala senega*—Snake root–roots
2. *Trigonella foenum graecum*—Fenugreek–seeds
3. *Foeniculum vulgare*—Fennel–fruits
4. *Allium sativum*—Garlic–Cloves
5. *Sesamum indicum*—Sesame–seeds
6. *Taraxacum officinale*—Dandelion–roots

16. Herbs to Avoid During Pregnancy

All the plants that are emmenagogue in nature or having any effect on uterine muscles or hormonal activity are to be avoided during pregnancy.

1. *Rhamnus frangula*—Alder buckthorn–cathartic
2. *Aloe barbadensis*—Aloe–cathartic
3. *Ephedra sinica*—Ma-huang–cardiac stimulant
4. *Tanacetum parthenium*—Feverfew–emmanagogue
5. *Hydrastis canadensis*—Goldenseal–uterine stimulant
6. *Ruta graveolens*—Rue–emmenagogue
7. *Salvia officinalis*—Sage–emmenagogue
8. *Cassia senna*—Senna–laxative
9. *Smilax* spp.—Sarasaparilla–hormonal reactions
10. *Artemisia absinthium*—Wormwood–Emmenagogue

Index

Abelmoschus esculentus 194
Abranin 119
Abrasine 119
Abrectorin 119
Abrine 119
Abrins 119
Abrus precatorius 119
Absinthe 83
Absinthin 83
Abuta 33
Abutilon 224
Abutilon avicenniae 196
Abutilon glaucum 196
Abutilon hirtum 196
Abutilon indicum 13, 195
Abutilon pannosum 196
Abutilon ramosum 196
Abutilon theophphrastii 196
Abyssinian Tea 13
Acacetin 184
Acacia catechu 186
Acacia nilotica subsp. *indica* 187
Acacia senegal 196
Acer tannin 181
Acetophenones 141
Acetyl choline 14
Acetylacetal 128
Achillea millefolium 83
Achilleine 84
Achyranthes aspera 101
Achyranthine 101
Aconine 14
Aconite 10, 14
Aconitines 14
Aconitum 14
Aconitum chasmantham 15
Aconitum ferox 15
Aconitum heterophyllum 15
Aconitum napellus 14
Acorin 62
Acorone 62
Acorus calamus 62, 225
Acrylylshikonin 97
Actaea racemosa 128
Actein 128
Acutangulic acid 182
Adenanthera pavonina 186
Adhatoda vasica 41
Adhatoda zeylanica 41, 114, 227
Adonis 205
Aegeline 147
Aegle marmalos 147
Aerva lanata 107, 228
Aescisn 122
Aesculus hippocastanum 122
Afim 34
Agar 201
Agaropectin 201
Agarose 201
Agavasaponins 120
Agave americana 120
Age-related macular degradation (AMD) 159
Ageratochromene 82
Ageratum conyzoides 82
Agnimanthah 53
Agnuside 79
Ailanthic acid 113
Ailanthione 113
Ailanthus excelsa 113
Ajmalicine 17
Ajoene 90
Akebia quinata DC 118
Akeboside 118
Akuammidine 24
Alangicide 30
Alangimarkine 30
Alangine 30

Alangium lamarkii 29
Alangium salvifolium 29
Alantolactones 87
Alatolide 58
Albaspidine 142
Albizzia lebbeck 187
Albizzia odoratissima 188
Albizzia procera 188
Alcea rosea 197
Alder Buckthorn 178
Aleurinins A,B 95
Aleurites fordii 95
Aleurites moluccana 154
Aleurite triloba 154
Alexandrian Laurel 98
Algin 200
Alginic acid 200
Alhagain 12
Alhagi camelorum 12
Alhagi maurorum 12
Alhagi manna 12
Alhagi pseudalhagi 12
Alkaloidal amines 11
Alkaloids 9
Alkarka 133
Allamanda cathartica 57, 101
Allamandin 101
Allicin 90
Alliin 90
Allium cepa 91
Allium porrum 91
Allium sativum 90, 225, 227, 228, 229
Allium schoenoprasum 91
Allo-imperatorin 147
Allyl isothiocyanate 88
Allyl propyl disulphide 90
Alocutin 174
Aloe barbadensis Mill. 173, 226, 229
Aloe vera 173
Aloe-emodin 176, 177, 178
Aloesin 174
Aloesone 174
Alpinia galanga 62
Alpinia officinarum 63
Alstonia scholaris 24
Althea officinalis 197
Althea rosea 197
Alzheimer's disease 2, 180, 190
Amalaki 209
Amaltas 176
Amanita 10
Amarogentin 57, 80, 94
Amaropanin 80
Amaroswerin 57, 80, 94
Ambrosia 58
Amentoflavone 156
American Aloe 120
American Wormseed 71
American Yellow Jasmine 26
Aminoacids 206
D-Aminoacids 207
α-Aminoadipic acid 207
α-Aminopimelic acid 207
Ammi majus 225
Ammi visnaga 149, 150
Amomum subulatum 63
Amomum zedoria 76
Amorphophallus campanulatus 199
Amruta 32
Amygdalin 205
Anacardic acid, 143
Anahygrine 112
Anamirta cocculus 86
Anantamool 38, 104
Anar 40
Andira araroba 179
Andirobin 112
Andrographis paniculata 95, 227
Andrographolide 95
Anethole 75
Anethum graveolens 63
Anethum sowa 64
Angelica archangelica 65
Angelica glauca 65
Angelica sinensis 60
Angelicain 65
Angustibalin 58
Anhydroscilliphaosidin 136
Anjir 110
Ankol 29
Annato 139
Anogeissus latifolia 194
Anomospermum grandiflora 32
Anthemic acid 82
Anthemis nobilis 81
Anthocercis littorea 52
Anthochlor pigments 156
Anthocyanins 155
Anthoxanthins 157
Anthraquinones 171
Anthrones 177
Antioxidant Therapy 189

Apamargah 101
Aparajitin 212
Apigenin 65
Apium graveolens 65, 225
Aporphines 27
Arabinogalactan 193
Arabinoxylans 193
Arborone 152
Archangelin 65
Arctiin 151
Arctium 151
Ardusa 113
Areca 10
Areca catechu 38
Areca nut 38
Arecaine 38
Arecoline 38
Argemone mexicana 28, 31, 225
Argyreia nervosa 20
Aristolactam 35
Aristolochia indica 35
Aristolochic acid 35
Arjun 185
Arjunetin 185
Arjunic acid 185
Arkah 132
Armoracia lapathifolia 88
Armoracia rusticana 88
Arnebia nobilis 171, 172
Arnebin 171
Arnica 58
Arnicolide 58
Arnidiol 66
Aromatherapy 59
Artemisetin 83
Artemisia 83
Artemisia absinthium 83, 229
Artemisia cina 82
Artemisia maritima 83
Artemisin 83
Aryl chromans 159
Asafoetida 91, 172
Asafoetidin 92
Asanah 168
Asaraldehyde 62
Asaresinol ferulate 92
Asarone 62
Ascaridole 71
Asclepiadin 131
Asclepias curassavica 130
L-Ascorbic acid 206
Ash Gourd 199
Ashta 165
Ashvaghna 135
Ashwagandha 111
Asiaticoside 126
Asoka 183
Asparagine 65
Asparagus adscendens 121, 224, 228
Asparagus racemosus 120, 121, 226
Aspergillum fumigatus 20
Aspidinol 142
Aspidium 142
Aspidosperma 57, 228
Asteracantha longifolia 193
Asthisandana 111
Asvasakhotah 43
Aswattah 110
Atherosclerosis 2, 190
Atibala 13, 195
Atidine 15
Atisine 15
Ativish 14
Atropa 7
Atropa belladonna 48
Atropine 11, 48, 51
Aurone 63
Aurones 156
Autumnolide 58
Avena sativa 228
Averhoa bilimbi 227
Azadirachta indica 114, 226
Azadirachtin 112, 114
Azadiradione 114
Azadirone 113
Azetidine-2-carboxylic acid 207
Azulene 65, 83

Babul 187
Bacchanag 14, 15, 53
Bacogenins 124
Bacopa monnieri 124, 126, 225, 227
Bacoside 124
Baicalein 163
Baileya 58
Bakrachimyaka 151
Bala 13, 224
Balanites aegyptiaca 123
Balanites roxburghii 123
Balanitisins A-E 123

Balduina 58
Baliospermin 96
Baliospermum axillare 96
Baliospermum montanum 96
Baliospermum solanifolium 96
Ballon Vine 167
Balsamic acids 146
Bamboo 198
Bambusa arundinacea 198
Bambusa bambos 198
Bangal Quince 147
Bangra 70
Bansa 198
Banslochan 198
Banyan Tree 109
Barakulanjar 62
Baramasi 17
Barbaloin 174, 178
Bari Elachi 63
Barlacristone 79
Barleria cristata 79
Barleria prionitis 78, 224
Barleria strigosa 79
Barlerin 78
Barringtogenin 182
Barringtogenol 182
Barringtogentin 182
Barringtonia acutangula 182
Barringtonic acid 182
Bartogenic acids 182
Basella alba 198
Basella rubra 198
Basil 76
Basseol 175
Bauhinia purpurea 165
Bauhinia racemosa 165
Bauhinia tomentosa 166
Bauhinia variegata 166
Bavchi 148
Bebeerine 33
Behada 185
Bel 147
Belladona 10, 48
Belliric Myrobalan 185
Bellyache bush 96
Bengal Kino 188
Bengalenoside 109
Benincasa cerifera 199
Benincasa hispida 199
Benzoin 145
Benzophenones 142
Benzoquinones 171
Benzyl Isoquinolines 27
Bis-Benzyl isoquinolines 27
Benzyl isothiocyanate 87, 88
Berberine 28, 30, 31
Berberis 28, 31
Berberis aristata 29
Berberis asiatica 29
Berberis valgaris 31
Bergaptene 64, 65, 147, 150
Bergenia ciliata 149
Bergenia ligulata 149
Bergenin 111, 147, 149, 170
Betacyanins 209
Betaine 12
Betel 76, 77
Betel nuts 39
Betula lenta 70
Bhadra 107
Bharangi 109
Bhat 167
Bhilawa 142
Bhilawanol 143
Bhoomyamalki 152
Bhringaraja 148
Biflavonyls 156
Bignonia indica 163
Bilberry 159
Bilobalide 93
Bilobals 93
Biochanin A 188
Biophytum sensitivum 163
Bisabolol 58
α-Bisacolol 82
Bixa orellana 139
Bixaghanene 139
Bixin 139
Bixorellin 139
Black Cohosh 128
Black Hellebore 136
Bloodroot 31
Boerhavia diffusa 44, 102, 224, 228
Boerhavia repens 44
Bonduc nut 98
Bonducellin 98
Bonesetter 111
Bornyl acetate 68
Boswellia serrata 97, 227
Boswellic acids 97

Brahmi 124, 126
Brahmine 124
Brahmoside 126
Brassinosteroids 175
Brazilin 159
Brindon 210
Brindonia indica 210
Broom Tops 43
Brucea amarissima 114
Brucea antidysenterica 113, 114
Brucea javanica 114
Bruceajavanin A 115
Bruceanols 115
Bruceantin 115
Bruceantinol 115
Bruceine 115
Bruceoside C 115
Brucicanthinoside 115
Brucine 11, 18
Bruguiera sexangula 52
Buckwheat 161
Buddhist Bauhinia 166
Bufadienolides 128
Bupleurum falcatum 117
Butea gum 188
Butea frondosa 188
Butea monsperma 188
Butein 189

Cadabacine 54
Cadavarine 10
Caesalpinia bonduc 98, 227
Caesalpinia coriaria 181
Caesalpinia crista 98
Caesalpinia nuga 98
Caesalpinia spinosa 181
Caesalpins 98
Caesulia axillaris 70
Caffeine 162
Calabar Bean 19
Calactin 134
Calandulosides 66
Calendula 66
Calendula officinalis 66, 226
Calophyllic acid 99
Calophyllin 99
Calophyllolide 99, 159
Calophyllum 159
Calophyllum inophyllum 98
Calotropain 132
Calotropin 132, 133, 134
Calotropis gigantea 132
Calotropis procera 133
Calumba Root 34
Calumbamine 34
Calumbin 34
Cambogin 211
Camboginol 211
Camel thorn 12
Camellia sinensis 162
Campanilla 101
Camphor 72
Camphor 62
α-Campholenic acid 97
α-Campholytic acid 97
Camptothecine 11
Cancer 2, 190
Canda 126
Candle Nut Tee 154
Cangeri 210
Cannabinoids 57
Canscora 167
Canscora decussata 167, 225
Canthin-6-one 107
Canthiumine 207
Capsaicin 144
Capsainoids 144
Capsanthin 144
Capsella bursa - pastoris 89
Capsicum annum 143, 228
Capsicum frutescens 143
Capsicum minimum 143
Capsicum purpureum 143
Capsorubin 144
Caraway 70
Carbohydrates 205
β-Carbolines 25
Cardamom 63
Cardenolides 128
Cardiac Glycosides 55, 128
Cardiospermum halicacabum 166, 227
Carica papaya 137
Carissa carandas 131
Carissol 131
Carissone 131
Carob 200
β-Carotene 137–139
Carotenoids 2, 55, 136
Carpaine 138
Carrageenin 201

Carum carvi 70
Carvone 64, 66, 70
l-Carvone 69
Cascara Bark 178
Cascarosides 178
Cassia angustifolia 177
Cassia fistula 176, 226
Cassia occidentalis 174, 226
Cassia senna 175, 229
Cassia tora 175, 226
Cassioside 72
Castor 213
Catechin 156, 186, 187, 200
Catha edulis 13
Catharanthus roseus 17, 225
Cathedulins 13
Cathidine 13
Cathinone 13
Cat's claw 18
Catunaregam spinosa 125
Celapagin 44
Celapanigin 44
Celastrine 45
Celastrus paniculatus 44, 226
Celery 65
Centella asiatica 126, 225, 226, 227
Cephaeline 29
Cephaelis ipecacuanha 28
Cephalotaxus harringtonia 54
Ceratonia 205
Ceratonia siliqua 200
Cerbera manghas 131
Cerbera odollum 131
Cereberoside 135
Cerleaside 132
Ceveratrum alkaloids 45
Chai 162
Chakramarda 175
Chalcone 63, 157
Chamazulenes 58
Chamomile 81
Chasmaconitine 15
Chasmanthinine 15
Chatinine 37
Chaulmoogra 213
Chaulmoogric acid 213
Chavibetol 77
Chavicine 39
Chebulagic acid 184, 185
Chebulic Myrobalan 184
Chebulinic acid 184
Chelerythrine 32, 42
Chenopodium ambrosioides 71
Chenopodium anthelminticum 71
Chestnut tannin 181
Chicoric acid 142
Chicory 84
Chinese Cassia 71
Chitrak 173
Chives 91
Chlorogenic acid, 142
Chlorophytum arundinaceum 121
Chlorophytum borivilianum 121, 224, 228
Chlorophytum tomentosum 121
Choline 89
Chondrodendron 32
Chondrodendron microphylla 32
Chondrodendron tomentosum 32
Chondrus 201
Chondrus crispus 201
Chotta Chirayata 37
Christ's Thorn 131
Chromenoflavone 169
Chromones 149
Chrysanthemum parthenium 80
Chrysanthemum vulgare 62
Chrysarobin 179
Chrysin 163
Chrysophanol 176
Chymopapain 138
Cichorium intybus 84
Cimifugoside 128
Cimigenol 128
Cimicifuga racemosa 128
Cinca 210
Cinchona 10, 41, 42
Cinchona calisaya 42
Cinchona ledgeriana 42
Cinchona officinalis 42
Cinchona succirubra 42
Cinchotannic acids 42
Cineole 61, 63, 68, 74
Cinnacassiols 71
Cinnamaldehyde 72
Cinnamic acid 78, 146
Cinnamomum aromaticum 71
Cinnamomum camphora 72
Cinnamomum cassia 71
Cinnamomum tamala 72

Cinnzeylanin 71
Cinochonine 42
Cissampelos pareira Linn. 33
Cissus quadrangula Linn. 111, 226
Citral 74
Citric acid 206
Citronellal 61
Citrullus colocynthis 112
Citrus 140, 208
Citrus aurantifolia 208
Citrus limon 208
Citrus medica 208
Citrus paradisi 208
Citrus reticulata 208
Citrus sinensis 208
Claviceps 10
Claviceps purpurea 20
Clepogenin 131
Clerodendrum indicum 109
Clerodendrum siphonanthus 109
Clitoria 212
Cnicin 58
Cnicus 58
Coca 51
Cocaine 11, 51
Coccinia grandis 225
Cochlearia arctica 52
Cochlospermum gossypium 194
Codeine 11, 28, 35
Coffee Senna 174
Colchiceine 52
Colchicine 52, 53
Colchicum autumnale 52
Coleus amboinicus 73, 227
Coleus aromaticus 73
Colocynth 112
Colocynthis vulgaris 112
Columbin 32
Commiphora abyssinica 106
Commiphora erythraea 106
Commiphora myrrha 106
Commiphora shimperi 106
Commiphora wightii 106, 227
Commiphoric acid 106
Coneflower 192
Conessine 46, 47
Coniine 40
Conium 10
Conium maculatum 40
Convallaria majalis 130
Convolvine 21
Convolvulin 214
Convolvulinic acid 214
Convolvulus arvensis 51
Convolvulus microphyllus 21
Convolvulus pluricaulis 21
Convolvulus prostratus 21, 225
Convolvulus pseudo-cantabrica 52
Conyza cinerea 102
Coralwood 186
Corchorus olitorius 133
Cordifolioside 33
Coriandrum sativum 73
Corilagin 181
Corkwood Tree 50
Corylus avellana 94
Corynanthe 57
Costunolide 58
Costus speciosus 126, 226
Cotton 224
Coumarino-lignoids 95, 104, 154
Coumarins 146
Country Mallow 195
Crab's eye 119
Crataegus laevigata DC 160
Crataegus monogyna 160
Crataeva nurvala 54
Crataegus oxycantha 160
Creat 95
Cristabarlone 79
Crocin 138, 139
Crocus sativus 138
Croton tiglium 93, 96
Cubeb 152
Cubebin 153
Cucurbitacins 100
Cularines 27
Curare 32
Curcolene 76
Curculigo orchioides 127, 224
Curculigo malabarica 127
Curcuma longa 144, 226, 227
Curcuma zedoaria 76, 226
Curcuma zerumbet 76
Curcumin 145
Curcuminoids 145
Curcumol 76
Cuscohygrine 48, 52
Cutch Tree 186
Cyanaropicrin 58

Cycloartenol 108
Cymbopogon citratus 74
Cymbopogon martinii 74
Cynara 58
Cyperotundone 87
Cyperus rotundus 87
Cytisus scoparius 43

Daemia extensa 134
Dalbergiones 159
Damar 83
Damsin 58
Dandelion 125
Danti 96
Daphne mezereum 92
Daruharidra 29
Dasamula 25, 53, 163, 173
Dasapushpam 107
Dasparsha 121
Datura ferox 52
Datura metel 50
Datura sanguinea 52
Datura stramonium 49, 50
Dead Sea Apple 133
Delphinium 14
Delphinium ajacis 16
Depsides 142
Derris indica 169, 226
Desert Date 123
Desmodium gangeticum 25
Dextrins 206
Dhanya 73
Dhataki 164
Dhatura 49
Dhawa 194
Diabetes 190
Diadzin 168
Dicentra 27
Dietary fibre 206
Digitalis 128
Digitalis lanata 129
Digitalis purpurea 128
Digitoxigenin 129, 134
Dihydromethysticin 145
Dikamali 164, 165
Dioscin 127
Diosgenin 120, 122
Dita bark tree 24
Diterpenes 92
Diterpenoid Alkaloids 14
Diterpenoids 55
Djenkolic acid 207
Dong quai 60
DOPA 19, 165, 207
Dravanti 96, 97
Dregianin 113
Drimia indica 136
Drumstick 89
Dryopteris filix-mas 142
Drypetes roxburghii 107
Duboisia myoporoides 50
Duboisia leichhardtii 50
Dugdhika 108
Dulcitol .206
Dumetorins 126
Duralabha 121

β-Ecdysone 44
Ecdysterone 13, 01, 102
Ecgonines 51
Echinacea angustifolia 192, 227
Echinocystic acid 188
Echitamidine 24
Echitamine 24
Eclipta alba 148, 227
Eclipta erecta 148
Eclipta prostrata 148
Ecliptal 148
α-Elaeostearic acid 95
Elatericin 112
Elaterin 112
Elderberry 160
Elemanolides 58
Elephant Foot Yam 199
Elephantin 58
Elephantopin 58
Elephantopus 58
Elephantopus scaber 86
Eleutherococcus senticosus 117
Eleutheroside 117
Ellagitannins 182
Embelia ribes 171, 172
Embelin 171, 172
Emblica officinalis 184, 209
Emetic Nut 125
Emetine 11, 28, 29, 36
Emodin 175, 178
Enicostemma hyssopifolium 37
Enicostemma littorale 37, 225
Ephedra 11, 140

Ephedra sinica 11, 228
Ephedrans 12
Ephedrine 12
Ephedrine 11, 13
Epigallocatechin 162
Equisetum arvense 228
Erandah 213
Eremantholide A 58
Eremanthus 58
Ergocristine 20
Ergot 20
Ergotamine 20
Ergotoxine 20
Ervatamia divaricata 26
Erysimoside 133
Erysodine 23
Erysotrine 23
Erythrina indica 23
Erythrina variegata 23
Erythrophleum 14
Erythroxylum coca 51
Erythroxylem traxillense 51
Eseramine 19
Estragol 148
Eucalyptus 7
Eucalyptus globulus 74
Eugenia aromatica 67
Eugenia caryophyllata 67, 149
Eugenia jambolana 183, 225
Eugenol 67, 72, 76
Euglobal 74
Eupatorin 58
Eupatoripicrin 58
Eupatorium 58
Euphol 109, 175
Euphorbia hirta 108
Euphorbia esula 93
Euphorbia neriifolia 108
Euphorbia nivulia 108
Euphorbia prostrata 108
Euphorbia pilulifera 108
Euphorbia thymifolia 108
Euphorbins 108
European Dill 63
Evening Primrose 212
Evolvulus alsinoides 21, 225, 228
Excelsin 113
Exogonium purga 213

Fagonone arabica 121
Fagonia cretica 121
Fagonone 121
Fagopyrum esculentum 161
False Pareira 33
Farnesiferol A 92
Farnesol 58
Featherfew 80
Female Ginseng 60
Fennel 74
Fenugreek 122
Feronia limonia 148
Ferula asafoetida 91
Fever nut 98
Feverfew 80
Ficus benghalensis 109, 226
Ficus carica 110
Ficus racemosa 110, 226
Ficus glomerata 110
Ficus religiosa 110, 172, 226
Filicin 142
Fish Berries 86
Flavan-3, 4-diols 156
Flavones 157
Flavonoids 57, 155
Flavonols 157
Flavonones 158
Flavononols 158
Flax 153
Foeniculosides 75
Foeniculum vulgare 74, 229
Folic acid 208
Frangula Bark 178
Frangulin B 178
Frankincense 97
Free radicals 189
Friedelin 107, 108
Funtumia elastica 47
Funtumine 47
Furanocoumarins 75

Gaillardia 58
Gaillardin 58
Galactomannan 200
Gamboge 211
Gangatulsi 77
Ganiarine 53
Ganikarine 53
Garcinia combogia 211
Garcinia indica 210, 227
Garcinia mangostana 211

Garcinia morella 211
Garcinia purpurea 210
Garcinol 211
Garcinones 211
Garden Nasturtium 88
Garden Rue 77
Gardenia gummifera 164
Gardenia resinifera 165
Gardenins 165
Garlic 90
Garrya 14
Gartenin 211
Garudaphala 213
Gaultheria procumbens 69
Gaultherin 70
Gedunin 112, 114
Gelidium cartilagineum 201
Gelsemicine 26
Gelsemine 26
Gelsemium sempervirens 26
Genistein 168
Gentian 80
Gentiana lutea Linn 80, 146
Gentianine 80
Gentianose 205
Gentiopicrin 80
Gentisin 146
Germacranolides 58
Germacrolides 86
Germichrysone 175
Germine 45
Giganteol 132
Gigantin 132
Gigartina mamillosa 201
Ginkgo biloba 93, 226, 228
Ginkgolides 93
Ginseng 116
Ginsenosides 116
Glabranin 123
Glaucarubin 113
Glinus oppositifolius 116
Gloriosa superba 53
Gluanol acetate 110
Gluanone 167
Glucosinolates 87
Glucotropaeolin 87, 88
Glutathione 145, 207, 209
Glutathione-S-transferase 59
Glycine max 167, 226
Glycine soja 167
Glycitein 168
Glycosides 55
Glycosmis pentaphylla 43
Glycyrrhetol 123
Glycyrrhiza glabra 123, 227
Glycyrrhizin 119, 123
Glycyrrhizinic acid 123
Gmelina arborea 152, 227
Gmelofuran 152
Gnidia lamprantha 92
Gnidin 92
Goa Powder 179
Goatweed 82
Gojiva 86
Gokhru 127
Goksura 127
Golden Seal 30
Gossypetin 195
Gossypium arboreum 103
Gossypium herbaceum 103, 224
Gossypol 103, 184
Gotu Kola 126
Gracilaria confervoides 201
Guaiacum 151
Guaiacum officinale 125
Guaianolides 58
Guaiazulene 58
Guaicins 125
Guggul 106
Guggulusterol 106
Gum Arabic 196
Gums 192
Gunja 119
Guttiferins 211
Gymnema 105
Gymnema sylvestre 105, 225
Gymnemagenins 105
Gymnemic acids 100, 105
Gymnestrogenin 105

H.spinosa T. Anders. 193
H_2O_2 191
Haematoxylin 183
Halenium 58
Halim 87
Hamamelis 181
Hamamelis virginiana 182
Hamamelitannin 181, 182
Haridra/Haldi 144
Harmal 25

Harmalol 25
Harman 25
Harmine 24, 25
Harringtonine 54
Harsinghar 139
Hawthorn 160
Hayatine 33
Hayatinine 33
Hecogenin 120, 121
Hedera helix 36
Hedera nepalensis 36
Hedra rhombea 36
Hederasaponins 36
Hederins 36
Hedysarum gangeticum 25
Heerabomyrrholic acids 106
Helicteres isora 103
Hellebore 45
Helleborus niger 136
Hellebrin 136
Hemedesmus indicus 104, 224
Hemiterpenoids 55
Henbane 49
Herpestris monnieri 124
Hibiscus esculentus 194
Hibiscus rosa—sinensis 160, 226
Hindisana 177
Hing 91
Hingoli 123
Hiranpag 51
Hirda 184
Histamine 14
Hoarhound 67
Holarrhena antidysenterica 46, 113
Holarrhena pubescens 46
Hollyhock 197
Holoptelia integrifolia 105, 227
Holoptelin A 105
Homeostasis 189
Homoisoflavones 158
Hops 143
Hordeine 12
Horse Chestnut 122
Horse radish 88
Huckleberry 159
Hulupones 143
Humulinones 143
Humulus lupulus 143
Hydnocarpic acid 213
Hydnocarpus 212
Hydnocarpus wightiana 213
Hydrangea 147
Hydrastis 10
Hydrastis canadensis 30, 229
Hydrocotyle asiatica 126
Hydrocotylene 79
Hydroxycitric acid 211
5-Hydroxy tryptamine 19
Hygrines 51
Hygrophila auriculata 193
Hygrophila spinosa 193, 224, 228
Hyoscine 49, 51
Hyoscyamine 11, 49
Hyoscyamus 10
Hyoscyamus muticus 49
Hyoscyamus niger 49
Hypaphorine 23
Hypericin 180
Hypericum perforatum 180, 226
Hyperin 180
Hypoglycin A 207
Hypophyllanthin 152
Hypoxis orchioides 127
Hyptis suaveolens 77
Hyssop 61, 206
Hyssopus officinalis 61

Iboga 57
Ichnocarpus frutescens 104, 224
Illudin 58
Imidazole Alkaloids 16
Indaconitine 15
Indian aconite 15
Indian Birthwort 35
Indian Borage 73
Indian Dill 64
Indian Ginseng 111
Indian Gooseberry 209
Indian Gum Arabic 187
Indian Laburnum 176
Indian Madder 179
Indian Napellus 15
Indian Oak 182
Indian podophyllum 151
Indian Sarasaparilla 104
Indian Sorrel 210
Indian Squill 136
Indigofera tinctoria 227
indocentelloside 126
Indole Alkaloids 16

Indravalli 167
Indravaruni 112
Ingenol 93
Insulin 206
Intellect Tree 44
Inula 14
Inula helenium 87
Inula racemosa 86
Inunolide 87
β-Ionones 66
Ipecac 10, 28
Ipecoside 29
Ipomoea digitata 22
Ipomoea hederacea 21
Ipomoea marginata 22
Ipomoea mauritiana 22
Ipomoea orizabensis 214
Ipomoea purpurea 20
Ipomoea turpethum 214
Iridoids 57, 78
Irilone 168
Iris germanica 168
Irish Moss 201
Irisolone 168
Isabgol 193
Ishwari 35
Ishwarone 35
Isobarbaloin 174
Isochondrodendrine 32
Isoflavones 158
Isoflavonones 158
Isomorellic acid 211
Isoquinoline Alkaloids 27
Isorottlerin 170
Ivy 36
Ixora coccinea 161

Jaborandi 16
Jalap 213
Jalapin 214
Jalapinolic acid 214
Jambuh 183
Japa 160
Jatamansi 81
Jatamansone 81
Jatamol 81
Jateorhiza palmata 34
Jatiphala 75
Jatoerhizine 34
Jatropha glandulifera 97
Jatropha gossypifolia 92, 96
Jatropha macrorhiza 92
Jatrophatrione 92
Jatropholones 96
Jatrophone 92, 96
Jawasa 12
Jerveratrum alkaloids 45
Jervine 45
Jew's mallow 133
Jivanti 135
Jungly Piaz 136
Jurinea 58
Jyotishmati 44

Kabab Chini 152
Kachanar 165, 166
Kaempferol 63, 64, 175, 177, 183
Kairata 94
Kakamari 86
Kakatundi 130
Kakmachi 46
Kala Dhatura 50
Kala Siris 188
Kala Zira 70
Kalimirich 39
Kalmegh 95
Kamal 36
Kamala 170
Kanchano 42
Kanphul 125
Kantakari 47
Kantala 120
Kapas 103
Kapoor 72
Karanj 169
Karanjah 105
Karaunda 131
Karavirah 134
Karaya Gum 194
Karcurah 76
Karpasa 103
Karpurvalli 73
Kasamarda 174
Kasani 84
Kasmari 152
Kattha 186
Kavach 19
Kavakava 145
Kesar 138
Khair 186

Khaira 197
Khat 13
Kheersal 186
Khella 150
Khellin 150
Khellinin 150
Khurasaniyajvayan 49
Kirmala 82, 83
Kokam 210
Kokilaksha 193
Koshmanda 199
Kota 148
Kotu 161
Kovidara 166
Krishnabijah 21
Kuberakshi 98
Kumari 173
Kundhandana 186
Kurchi 46
Kurcholessine 46
Kurutakah 134
Kutaja 47

Lactuca virosa 52
Lahsuna 90
Lajjalu 163
Lakshmana 22
Lalmirch, 143
Lanatosides 129
Lavandula officinalis 68
Lavender 68
Lebbekanin 187
Lectins 119, 208
Leek 91
Lemon Balm 61
Lemongrass 74
Lepachol 173
Lepidium sativum 87
Leptadenia reticulata 135
Liatrin 58
Liatris 58
Lignans 150
Lignum vittae 125
Ligustilide 61
Lilichai 74
Limbic system 60
Limonene 64, 65, 68, 70
Limonin 113, 208
Limonoids 7, 100, 112
Linalool 68, 73
Linamarin 153
Linoleic acid 212
Linseed 153
Linum album 151
Linum usitatissimum 153
Liquiritigenin 169
Liquorice 123
Liriodendrin 44
Liriodendron 58
Lobelia inflata 40
Lobeline 41
Lobinaline 41
Lochnera rosea 17
Locust Bean 200
Lodhra 185
Loliolide 66, 167
Long Pepper 39
Lonika 209
Lopeztree 42
Lophophora williamsii 27
Lotus 36
Low-density lipoproteins (LDL) 190
LSD 20
Luckmuna 50
Lumicolchicines 53
Lunularic acid 150
Lupinane Alkaloids 43
Lupulones 143
Luteolin 65
Lycopene 137
Lycopersicon esculentum 137
Lycopersicum 48
Lycorine 127
Lyonia–xyloside 151
Lysine 10

Ma-Huang 11
Machaerinic acid 188
Maclurin 211
Macrocystis pyrifera 200
Macrocystis 200
Madana 125
Madecassocide 126
Madhunashi 105
Madhurika 74
Magnolia Wine 154
Mahabala 13
Mahonia aquifolium 31
Maidenhair Tree 93
Makoi 46

Malabar Kino 168
Malabar Nut 41
Malatyamine 103
Malkagunin 45
Malkangani 44
Mallotus philippensis 170
Malvalic acid 176
Mammea americana 146
Mandragora officinarum 50
Mandrake 50, 151
Manduk parni 126
Mangifera 146
Mangiferin 146
Mangishta 179
Mangosteen 211
Mangostin 211
Mangusta 211
Mannitol 125, 205
Marking Nut 142
Marmelosin 57, 147, 150
Marrubium vulgare 67
Marsdenia tenacissima 224
Marshmallow 197
Matricaria recutita 82, 226
May-Apple 50, 151
Maytensine 54
Maytensinoids 54
Maytenus buchananii 54
Maytenus ovatus 54
Melanoxetin 187
Melissa officinalis 61
Metha piperita 68, 228
Mentha pulegium 69
Mentha spicata 69
Menthol 68
Menthone 68
Mesua ferrea 99
Methika 122
Methyl salicylate 70
Methylcinnamate 63
Methylnonyl ketone 77
Mevalonic acid 55
Mexican Poppy 31
Mezerein 92
Milk Thistle 170
Milkweed 132
Milkwort 115
Millfoil 83
Mimosa pudica 226
Mixed terpenoids 57
Molephantin 86
Molephantinin 86
Moluccanin 154
Momordica charantia 225
Monoterpenes 55
Monoterpenoid Alkaloids 37
Monoterpenoids 55
Morellic acid 211
Moringa oleifera 89, 227
Moringa pterygosperma 89
Moringine 90
Moringinine 90
Morphine 11, 35
Mountain Ebony 166
Mrigashinga 103
Mucilage 13, 192
Mucuna 19
Mucuna pruriens 19, 228
Mucunine 19
Multiradiatin 58
Mumiri 89
Munjistin 179
Murraya exotica 26
Murraya koenigii 26
Murrayanine 26
Musali 127
Musta 87
Myo-Inositol 206
Myricetin 183
Myristica 151
Myristica fragrans 75
Myristicin 64, 75
Myrrh 106

Nagabala 13
Nagar Moth 87
Nagchampa 98
Nagilactones 92
Nagkesar 99
Nahagenin 121
Naphthaphenanthridines 28
Naphthaquinones 171
Narcotine 28
Nardostachys jatamansi 81
Neem 114
Nelumbo nelumbo 36
Nelumbo nucifera 36
Nelumbo speciosum 36
Neo-quassin 113
Nerifloiol 109

Neriifolin 132, 135
Nerium indicum 134
Nerium odorum 134
Nerium oleander 134
Nilgiri 74
Nimba 114
Nimbidin 114
Nimbin 112, 114
Nirgundi 79
Nishindine 79
Non-protein amino acids 186
Nornuciferine 36
Nortriterpenoids 112
Nuciferine 36
Nutmeg 75
Nux vomica 18
Nuxvomica 10
Nyagrodhah 109
Nyctanthes arbor-tristis 139

Occidentalins 175
Ocimum sanctum 76
Ocimum tenuiflorum 76, 225, 227
Odollum 131
Odoratissimin 188
Odoroside 131, 134
Oenothera biennis 212, 225
Okanin 187
Okra 194
Oldenlandia corymbosa 228
Oleander 134
Oleanolic acid 182
Olitoriside 133
Olivil 132
Onion 91
Onjisaponins 118
Onocer-7-ene 111
Onopordon acanthium 58
Onopordopicrin 58
Operculina turpethum 214
Opium Poppy 34
Organic acids 2, 206
Orizaba Jalap 214
Ornithine 10, 48
Oroxylin-A 163
Oroxylum 163
Oroxylum indicum 163
Ostruthol 65
Otobain 151
Ovatifolin 58
Oxalis corniculata 210
Oxalis sensitiva 163
Oxindole alkaloids 18

Pacific Yew 94
Paclitaxel 94
Pala Indigo 47
Palas 188
Palash 184
Palasitrin 189
Palasonin 189
Palmatine 34
Panax ginseng 116, 224, 228
Panax quinque folius 116
Paniculatin 45
Pantothenic acid 208
Papain 138
Papaver 7, 28
Papaver somniferm 27, 34
Papaya 137
Papeeta 137
Paranti 161
Paribhadrah 23
Parisa 184
Parkinson's diseases 2, 190
Parthenolide 81
Pashanbheda 149
Passiflora incarnata 24
Passionflower 24
Patala 173
patchoulenone 87
Patisa 15
Peganum harmala 25
Pelletierine 11, 40
α-Peltatins 150
β-Peltatins 150
Penicillium chermesinum 20
Penicillium rroquefortii 20
Penniclavine 20
Pentanortriterpenoids 113
Pepper 39
Peppermint 68
Pergularia daemia 134
Pergularia extensa 134
Periploca indica 104
Periwinkle 17
Petasites hybridus 58
Petrocladia sp. 201
Petroselinic acid 64
Petroselinium crispum 228
Petunidin 63

Peucedanum graveolens 63
Peucedanum sowa 64
Phenanthro-indolizidine Alkaloids 37
Phenolic Acids 141
Phenolics 7, 140
Phenyl Propanes 141
Phenylalanine 10
Phlobaphenes 157, 180, 181
Phloroglucinol 142
Phorbol 93, 96
Phthalide Isoquinolines 28
Phyllanthin 152
Phyllanthus amarus 152, 228
Phyllanthus discoides 52
Phyllanthus emblica 209, 224, 227
Phyllanthus niruri 152
Phyllemblic acids 209
Physochlaina praealta 52
Physostigma venenosum 19
Physostigmine 11, 19
Phytohaemagglutinins 208
Picrasma excelsa 113
Picrorhiza kurroa 80
Picrotoxin 86
Pilocarpidine 16
Pilocarpine 11, 16
Pilocarpus 16
Pilocarpus jaborandi 16
Pilocarpus microphyllus 16
Pilocarpus pinnatifolius 16
Pilosine 16
α-Pinene 65
Pipal 39
Pipecolic acid 207
Piper betel 76
Piper cubeba 152
Piper longum 39, 172, 227
Piper methysticum 145
Piper nigrum Linn. 39, 227
Piyaz 91
Plantago ispaghula 193
Plantago ovata 193
Platicodins 118
Platycodon 118
Platycodon grandiflorum 118
Plumbagin 173
Plumbago coccinea 173
Plumbago indica 173
Plumbago rosea 173
Plumbago zeylanica 226
Plumericin 101
Plumieride 101
Podocarpus 92
Podocarpus gracilior 92
Podolide 92
Podophyllotoxin 150, 151
Podophyllum 150
Podophyllum emodi 151
Podophyllum peltatum 151
Poi 198
Poison Hemlock 40
Polygala senega 115, 229
Polygala tenuifolia 118
Polyphenols 2
Polyterpenoids 55
Pomegranate 10, 40
pongachromene 169
Pongamia glabra 169
Pongamia pinnata 105, 169
Pongapin 169
Populnetin 184
Populnin 184
Portia Tree 184
Portulaca oleracea 209
Prangolarin 65
Prasniparni 25
Precatorine 119
Premna corymbosa 53
Premna integrifolia 53
Premna obtusifolia 53
Premnine 53
Presenegenin 116
Prickly-Chaff Flower 101
Primary Metabolites 204
Proanthocyanidins 156
Prolapsus ani. 82
Proscillaridin A 136
Prostaglandins 108
Protein tyrosine kinase (PTK) 168
Proteins 206
Protopine 28
Pseudaconitine 15
Pseudoguaianolides 58
Pseudopurpurin 179
Psoralea corylifolia 148, 226
Psoralen 65, 146, 148
Psychotria ipecacuanha 28
Psychotrine 29
Psyllium 193, 205
Pterocarposide 169

Pterocarpus marsupium 168, 226
Pterocarpus santalinus 179, 226
Pterostilbene 150, 180
Pterygospermin 90
Pudina 68
Punarnava 44, 102
Punarnavine 44
Punarnavoside 44
Punica granatum 40
Purpurin 179
Purslane 209
Pushkarmula 86, 168
Putikaranjah 105
Putrainjivoside 107
Putranjiva 107
Putranjiva roxburghii 107
Putranjivadione 107
Pyrethrins 57
Pyrethrum parthenium 80
Pyridine-piperidine Alkaloids 38

Quassia 113
Quassia amara 113
Quassin 113
Quassinoids 112, 113
Quebraco 228
Quercetin 63, 64, 180, 183
Quercus macrolepis 181
Quinazoline Alkaloids 41
Quinazolines 13
Quinine 11, 42
Quinoline Alkaloids 41
Quinolizidines 43
Quinones 171

Raktachandan 179, 180
Rakthamachi 137
Randia 206
Randia dumetorum 125
Randia spinosa 125
Randialic acid 125
Randioside 126
Ratanjot 172
Rauwolfia 17
Rauwolfia serpentina 23, 225
Rauwolfia tetraphylla 24
Rauwolfia vomiforia 24
Red Sandalwood 179
Remijia pedunculata 42
Rescinnamine 23
Reserpine 11, 17, 23
Resins 55
Rhamnus alnus 178
Rhamnus frangula 229
Rhamnus purshiana 178
Rhein 175, 176, 177
Rheum palmatum 177
Rhizopus arrhizus 20
Rhubarb 177
Rhus semialata 181
Ribitol 205
Ricin 213
Ricinine 213
Ricinoleic acid 212, 213
Ricinus communis 213, 224
Rivea corymbosa 20
Roemerine 36
ROS (Reactive Oxygen Species) 190
Rosemarinus officinalis 68
Rosemary 68
Rotenoids 169
Rotenone 57
Rottlerin 170
Roxburgholone 107
Rubber and gutta 55
Rubia cordifolia 179, 224, 225, 226
Ruta graveolens 77, 229
Rutin 12, 180

Sabal serrulata 124
Sabinene 77
Safed Chandan 84
Safed musali 121, 224
Safed Siris 188
Saffron 138
Safrole 61, 182
Sahachara 78, 224
Sahadevi 82, 102
Saikosaponins 118
Salai 97
Salix 227
Salsodine 12
Salvia officinalis 229
Sambucus nigra Linn 160, 228
Samudraphal 182
Sandalwood 84
Sanguinaria canadensis 31
Sanguinarine 32
Santalol 84
Santalum 150

Santalum album 84
Santamarine 81
Santene 84
Santonin 83
Saponins 55, 100, 101, 115
Saptaparna 24
Saraca asoca 183
Saraca indica 183, 226
Sarapunkhah 169
Sarasaparilla 116, 229
Sarasapogenin 116, 120
Sariva 104
Sarpagandha 23
Saunf 74
Saw Palmetto 124
Schisandra chinensis 151, 154
Schisandrin 154
Schizandrin 151
Scoparia dulcis 228
Scoparius 43
Scopolamine 48, 49
Scopoletin 48, 64, 66
Scopolia carniolica 50
Scutiamine 208
Secoisolariciresinol 153
Secondary Metabolites 2, 7
Sedanolide 65
Semecarpus anacardium 142
Senegenin 115
Senna 175
Sennidin 176, 177
sennoside 175, 176
Serenoa repens 124, 228
Serotonin 207
Serpentine 17
Sesame 153
Sesamol 153
Sesamolin 153
Sesamum indicum 153, 229
Sesquiterpene Alkaloids 44
Sesquiterpenes 57, 80
Sesquiterpenoids 55
Sesterpenoids 55
Sesuvin 102
Shallaki 97
Shankhapushpi 21, 167
Shatavarins 120
Shataveri 120
Shepherd's Purse 89
Shoe flower 160
Shonapushpe 139
Shyonaka 163
Siberian Ginseng 117
Sickle Senna 175
Sida 224
Sida acuta 13
Sida cordata 13
Sida cordifolia 13, 226, 228
Sida rhombifolia
Sida spinosa 13, 224
Sigru 89
Silybin 170
Silybium marianum 170, 228
Silychristin 170
Silymarin 170
Siris 187
Sitoindoside-X 13
β-Sitosterol 64, 66, 175, 182, 183
Skimmianine 43
Slippery Elm 196
Smilagenin 116
Smilax 116, 229
Smilax aristolochiarfolia 110
Smilax febrifuga 116
Smilax regelii 116
Snakeroot 115
Soja hispida 167
Solamargine 47
Solanine 46
Solanum 48
Solanum americanum 46
Solanum dulcamera 48
Solanum incertum 46
Solanum nigrum 46
Solanum surattense 47
Solanum xanthocarpum 47
Solasodine 46, 48
Solidago virgaurea 228
Sonamukhi 175
Sorbitol 206
Sowa 64
Soyasapogenol 167
Soybean 167
Sparteine 43
Spearmint 69
Spermine alkaloids 12
α-Spinasterol 188
Spirojatamol 81
Spreading Hogweed 44
St. Johns' Wort 180

Stephania glabra 33
Sterculia spp. 194
Stereospermum suaveolens 173
Steroidal Alkaloids 45
Steroids 55, 100, 101
Sterolins 100
Stevioside 205
Stilbenes 150
Stinging nettle 13
Stroke 190
Strophanthin 130
Strophanthus hispidus 130
Strophanthus kombe 130
Strychnine 11, 18
Strychnos 32
Strychnos castelnaea 32
Strychnos crevauxii 32
Strychnos nux-vomica 18
Strychnos toxifera 32
Styrax spp 145
Styrax benzoin 145
Styrax paralleloneurus 145
Styryl pyrones 145
Subulaurone 63
Supari 38
Suran 199
Sweet Flag 62
Swernakshiri 31
Swertia chirata 94
Swertiamarin 37
Swertioside 37
Sweta musali 121
Swetadurva 83
Symplocos cochinchinensis 185, 226
Symplocos crataegoides 186
Symplocos paniculata 186
Syringaresinol 44
Syzygium aromaticum 67
Syzygium cumini 183
Syzygium jambolanum 183

Taalmakhana 193
Tabashir 198
Tabernaemontana coronaria 26
Tabernaemontana divaricata 26
Tagitinin 58
Tamalpathra 72
Tamarind 210
Tamarindus indicus 206, 210
Tambulah 76
Tanacetum parthenium 80, 229
Tanacetum vulgare 62
Tangulic acid 182
Tansy 62
Tara tannin 181
Taraxacerin 125
Taraxacum officinale 125, 229
Taraxasterol 66, 125
Tartaric acid 210
Taxodione 92
Taxodium distichum 92
Taxodone 92
Taxol 94
Taxomyces andreanae 94
Taxus baccata 94
Taxus brevifolia 94
Taxus chinensis 94
Tea 225
Tecomine 37
Tephrosia purpurea 169
Terminalia arjuna 185, 225
Terminalia bellirica 184, 185, 227
Terminalia chebula 181, 184, 228
Terpenoids 55
Terrestiamide 127
Terrestrosins 127
Tetranortriterpenoids 112
Tetraterpenoids 55
Thalictrum minus 27
Thankuniside 126
Theaflavins 162
Theafolisaponin 162
Thearubigins 162
Theasapogenols 163
Theasinensins 162
Thespesia populnea Sol. 184, 228
Thevetia nerifolia 135
Thevetia peruviana 135
Thiophenes 148
Thorn Apple 49
α-Thujene 97
Thujone 62, 83
Thujyl alcohol 83
Thyme 69
Thymol 69
Thymus vulgaris 69
Til 153
Tiliacora 34
Tiliacora racemosa 34
Tiliacorine 34
Tinocordifolioside 33

Tinospora cordifolia 32, 224, 225, 226
Tinosporaside 32
Tithonia 58
Tomatine 137
Tomato 137
Trachyspermum ammi 227
Tree of heaven 113
Trianthema monogyna 102
Trianthema portulacastrum 102
Trianthemine 102
Tribulus terrestris 127, 228
Tricoccin S22 112
Trigonella foenum-graecum 122, 225, 229
Trilobolide 58
Tripdiolide 92
Triphala 184, 185
Tripterygium wilfordii 92
Triptolide 92
Triterpenes 101
Triterpenoids 55, 99
Trivert 214
Tropaeolum majus 88
Tropane Alkaloids 48
Tropanes 51
Tropolone Alkaloids 52
Tropolones 56, 57
Tryptophan 10
TS-polysaccharide 206
Tubocurarine 11, 32
Tulsi 76
Tung Oil 95
Turmeric 144
Turpethin 214
Tylophora 37
Tylophora asthmatica 38
Tylophora indica 38, 227
Tyrosine 10

Udumber 110
Ulmus rubra Muhl. 196
Umbelliferone 64
Uncaria guianensis 18
Uncaria tomentosa 18
Uncarine 18
Upodika 198
Urena 224
Urena lobata 13
Urginea indica 136
Ursina-Gurgi 211
Ursinia 58
Ursiniolide 58
Urtica dioica 13

Vaccinium myrtillus 159, 225
Vacha 62
Valerian 37
Valeriana 57
Valeriana officinalis 37
Valerine 37
Varuna 54
Vasaka 41
Vasakin 41
Vascinol 13
Vasicine 13, 41
Vasicinol 41
Vasicinone 13, 41
Vavding 172
veratramine 45
Veratrum viride 45
Vernolepin 58
Vernomenin 58
Vernomygdin 58
Vernonia 58
Vernonia amygdalina 58
Vernonia cinerea 102, 228
Vibhitaki 185
Vidarikanda 22
Vilayati piaz 91
Vinblastine 17
Vinca 17
Vinca pusilla 17
Vinca rosea 17
Vincarodine 17
Vincristine 17
Vishnukranta 21
Visnaga 150
Visnagin 150
Vitamin A 160
Vitamin C 206, 209
Vitamin E 212
Vitamin P 209
Vitex negundo 79, 225, 227
Vitis quadrangula 111
Volatile oils 2, 55, 59

Wedelolactone 148
Wild Indigo 169
Willow 227
Wintergreen 69

Witch Hazel 182
Withaferin 111
Withania somnifera 111, 225, 228
Withanolide 49, 111
Withanone 111
Withasomnine 112
Withastramonolide 49
Wood Apple 148
Woodfordia floribunda 164, 226
Woodfordia fruticosa 164
Wormseed 82, 83
Wormwood 229
Wrightia arborea 47
Wrightia tinctoria 47
Wrightia tomentosa 47
Wu-wei-zi 154

Xanthones 146
Xanthopurpurin 179
Xanthotoxin 65
Xeromphis spinosa 125

Ya-Danzi 114
Yahtimadhuh 123
Yohimbine 11
Yuccagenin 127

Zaluzania 58
Zaluzanin 58
Zedoary 76
Zingiber officinale 224–228
Ziziphine 208
Ziziphus jujuba 118